FOOD
and
POVERTY

£19-25.

FOOD
and
POVERTY

India's Half Won Battle

Gilbert Étienne

Sage Publications
New Delhi/Newbury Park/London

First published in 1988 by

Sage Publications India Pvt Ltd
32, M Block Market, Greater Kailash I
New Delhi 110 048

Sage Publications Inc
2111 West Hillcrest Drive
Newbury Park, California 91320

Sage Publications Ltd
28 Banner Street
London EC1Y 8QE

Published by Tejeshwar Singh for Sage Publications India Pvt Ltd., phototypeset by Printers Plates, Madras and printed at Chaman Offset Printers, New Delhi.

Library of Congress Cataloging in Publication Data

Étienne, Gilbert
Food and Poverty: India's Half Won Battle.
p. cm.
Bibliography: p.
1. Rural poor-India. 2. Food supply-India. 3. India-Rural conditions.
4. Agriculture-Economic aspects-India I. Title
HC 440. P6E86 1987 338.1'0954-dc19 87-16992
ISBN 0–8039–9540–7(US-hbk) 81-7036–076-5(India-hbk)
0–8039–9541–5(US-pbk) 81-7036–077-3(India-pbk)

Contents

Preface

Forty years after Independence, India has won two battles. The country is no more at the mercy of foreign grain and possibly international charity. Thanks to the reserves stored by the government, India is also free—and this *for the first time in its history*—from the danger of devastating famines.

A third battle, and this one is far from over, is that against poverty. One cannot deny that positive results have been achieved, especially in advanced districts, but a lot remains to be done.

While the struggle against poverty is playing an increasing role in development strategy, techno-economic issues remain considerably important. From that angle, one could say that India is now half way on the road to development. New inputs have been successfully introduced in many parts of the country. A lot of experience has been accumulated through trial and error so that today, overall policies can avoid several of the pitfalls which were unavoidable in the early phases of planning. On the other hand, agriculture has become increasingly demanding in terms of efficiency, sound planning and adequate supply of inputs.

What I call the 'muddle through system', i.e., an imperfect system where irrigation does not operate smoothly (temporary shortage of water due to power breakdown, poor maintenance of canals, etc.), where there is an inadequate supply of chemical fertilisers, late renewal of seeds, insufficient pest control, can, nevertheless bring considerable progress in poor but potentially rich areas like the eastern plains where such inputs have until now played a minor role. In advanced districts, where yields are already rather high, the 'muddle through system' will soon block further increases in production. There it has to be replaced by a system operating in a smooth and very efficient manner. After all, in order to approach European wheat yields and Japanese rice yields one cannot but ensure that all key inputs operate as they do in those countries.

Then comes a third category of regions which suffer from severe physical constraints in terms of soils, rains and irrigation, where the room for manoeuver is not so wide.

For all these reasons and bearing in mind the population pressure, the next steps of Indian agriculture are not going to be easy. Besides they will be more and more interdependent with the overall planning of the national economy in terms of research, of supply of inputs, of prices, of subsidies.

This book revolves around two basic concepts: time and space. Coming by car from Europe in October 1985, we made, my wife and I (until the end of February 1986), another round of surveys of villages and districts studied on several occasions since 1963. The areas selected provide us with a sample of several of the most typical situations found in India.

Listening to farmers, walking around the paddy fields, observing the people and the crops, collecting data from block and district officers and interviewing local officials have been my main activities. Such observations were then set into a broader framework, at the state level and finally, in New Delhi.

The timing of this research needs to be emphasised. The overall changes in the economic policy and the retreat of rhetorics[1] in favour of a more down-to-earth approach are of crucial importance not only for the country, but particularly for rural people.

Three basic issues are perfectly clear today. The already advanced areas, particularly Punjab, cannot feed India for ever; hence the growing importance of and concern for the sluggish eastern plains and their large untapped potential. Then what can be done in other—but for different reasons—slow moving areas especially in peninsular rainfed India? In Delhi some people claim that a breakthrough in such areas is essential, but this is more easily said than achieved.

Global issues are no less debated: how far should one advocate a direct attack on poverty through special schemes for the poor (target group approach) ? How far should one rely on the indirect approach whereby the overall growth of an area positively affects the lot of the poor? What are the major socio-political and techno-economic constraints? How can one envisage the future of rural India by the turn of the century? These are the main questions to be tackled in this book.

Chêne – Bougeries, Geneva Gilbert Étienne
May 1987

[1] Similar trends are noticeable in an increasing number of developing countries, particularly in China.

Acknowledgements

This survey of India's rural development has been made possible by a sabbatical from the Graduate Institute of International Studies and from the Institute of Development Studies in Geneva. The Swiss National Fund for Scientific Research helped me with a grant for part of my travel expenditure.

Such studies required some support from the Government of India. I would like to warmly thank Romesh Bhandari, the then Foreign Secretary and Suryakanti Tripati and Reena Poovaiah from his Ministry, for introducing me to the authorities of the states I visited. I cannot mention here all the officials whom I met in the field, at the district headquarters, in state capitals and in New Delhi. They were very helpful and, as is usual in India, very frank in expressing their views.

During the many weeks of driving through the country or walking around the villages, several of my old and dear friends were most generous in their hospitality. In New Delhi Vimal and Raghu Raj, Raksha and Pran Talwar, in Bombay Simone and Naval Tata, in Madras Thomas and Meera Abraham. We enjoyed staying with Surendra Pal Singh in Delhi and his son Ravin and his daughter-in-law Radesh, at Unchagaon.

As in my previous surveys, it has been rewarding to meet farmers and learn about their problems from them. With reference to Mirpur, particular thanks go to Ram Kumar Sharma, who supplied the data on his village.

While writing this book back in Geneva, I benefited from the comments of Jayante Prasad and his wife Bhanti, who belong to the Foreign Service and the Indian Administrative Service, respectively and Jean-Luc Maurer, Assistant Professor at the Institute of Development Studies. My thanks also go to those at the Institute who typed the manuscript.

Once again, my wife and I have made the long trip by car from Europe to India and back, getting away from our routine in Geneva. My work in the field and the writing of this book owe much to her. Finally, although I have had a great deal of support from many people, I alone am responsible for the facts and conclusions presented in this book.

Gilbert Éteinne

Chapter One

How to Approach the Rural Scene

Before coming to villages, wheat and paddy fields, let us focus briefly on the main topics to be studied and how to approach them.

For the distant past as well as for the contemporary world, one can talk of a *global process of rural development.*[1] Foodgrain or some other major crop makes a breakthrough or at least undergoes a sizable growth. Then other activities expand: secondary crops, milk, vegetables, orchards, fishing, etc. Trade and transportation, rural industries and, in modern times, rural electrification, play an increasing role. The links between the village and the town get stronger. These factors spring out through a chain reaction or, in a more or less simultaneous way, but they are all interrelated. Under their impact production increases and the local economy becomes more diversified.

For these reasons, one cannot confine observations to agriculture. No less attention has to be given to other aspects of the rural economy within the village and in the bigger villages or small towns. As will be shown in this book, when the global process of rural development is strong enough, *vis à vis* population growth, it does have a positive impact on the poor in terms of wages and job opportunities.

Several other variables of a non-economic nature also enter the picture. First is the increasing pressure of population on land and on employment. Then come political factors, from local politics and village *netas,*[2] to the broader issues at the level of the state or the centre. They may affect the development process in a positive or negative way.

[1] In my book *Rural Development in Asia* (New Delhi, Sage 1985), I have explained the similarities of this global process in medieval Europe and old China as well as in several contemporary Asian countries.

[2] Local leaders

A variety of approaches

Three main approaches are followed to ascertain what is happening in the agricultural and rural economy as a whole. The central government collects data from the states to bring out the broad picture, i.e., production trends crop-wise, consumption of inputs, increase in irrigation. Possible imports and exports of agricultural commodities are no less important to confirm production trends. Agricultural censuses also supply us with a lot of information.

The second approach involves research using data collected on a limited, but relatively large scale, such as that supplied by the numerous reports of the National Sample Survey, especially on conditions of living: employment, income, land tenure, etc. One also comes across studies conducted on a smaller scale in one state or the other, under research projects, using a number of field assistants.

In the third category fall micro studies such as village studies usually conducted by an anthropologist.

Each of these approaches has its own advantages and weaknesses. The first two, dealing with the macro-level, help to assess the main trends of the economy, bearing in mind that data cannot be completely reliable for lack of precision or for omission of certain items and sources of income. For instance, the poorer a man, the more difficult it is to measure his income which can be derived from several sources, themselves varying in course of time.[3]

Gobal data can also be misleading when different areas and people are amalgamated under the same head. For instance, data on average yields and wages will be meaningless if they lump together an area of low yield and low wages with the neighbouring area where yield and wages can be twice or three times as high. In fact, regional studies which are needed so badly, especially in such a large and diverse country as India, must refer to relatively homogeneous zones. It can be a district, for instance in the Ganges basin, or a sub-district (*taluk* or *tehsil*) when the whole district is too diverse, covering widely different areas.

[3] For more details see L. Dumont, *Une sous-caste de 1 'Inde du Sud* (Paris, Mouton 1957); and, M.N. Srinivas, Village Studies, Participant Observation and Social Science Research in India, *Economic and Political Weekly*, Special Number, August 1975. I have dealt in greater detail with these problems in my book *India's Changing Rural Scene* (New Delhi, Oxford University Press, 1982).

As for village studies, especially when they are comprehensive enough, i.e, including both social and economic information,[4] they are of great value as they give us a detailed view of what is really happening in a particular place. Now the trouble with that approach is that first, there are not too many good village studies available. It seems that today even among certain anthropological circles, the village study is falling out of fashion. Second, we also need wider observations beyond the village.

A fourth approach has been followed by S.R. Sen and his team in a report prepared for the Reserve Bank of India. They have presented a most remarkable study of eastern India, involving valuable data and a detailed analysis of the different areas of that vast region. Such a study is of considerable help for more limited studies such as my own. It would be valuable if all the broad regions of India were covered in such a way and so competently.[5]

Another possible approach

To try and overcome the difficulties faced in the above approaches, I have resorted to the following method: I selected one particular village[6] for a study, as detailed as possible, of the main socio-economic forces at work: land-man ratio, production trends, wages, employment, population increase, land tenure, behaviour of cultivators, landless labourers, etc. Then, I gathered data on the block to which the village belongs and on the district and paid brief visits to some other places, Thus, one can be sure that the selected village is reasonably representatvie of the area. At the same time one can assess how government policy operates within a district and the impact of the various development agencies and departments.

[4] See the exhaustive study by J.L. Chambard, *Atlas d'un Village Indien* (Paris, Mouton, 1980), which covers all aspects of socio-cultural and economic life of a village in Shivpuri district, Madhya Pradesh, from 1957 to 1980.

[5] Reserve Bank of India, *Report of the Committee on Agricultural Productivity in Eastern India*, 1984.

[6] The study of Mirpur (Bulandshahr) is more detailed than the others. That is where I started in 1963 to get used to village studies, spending five months during my first visit. The first round elsewhere was more brief varying from a few weeks to nearly a month. I also revisited Mirpur more often. Besides, in Mirpur I had the help of Ram Kumar Sharma who collected a substantial amount of data for me.

Later, brief visits to the state capital enabled me to interview senior officers concerned with rural development and to study the state plan and other basic reports.

Now how to select the areas? India can be divided into three very broad categories of areas: *advanced districts,* mostly in the north-west and in the south-east; *poor but potentially rich areas,* such as the middle and lower Ganges basin to which one can add the plains of Assam and Orissa; and, *peninsular India,* where severe physical or geographical constraints limit the future growth of agriculture.

In the first category I have chosen three districts: Bulandshahr in western UP, Guntur in Andhra Pradesh, Thanjavur in Tamil Nadu. In the second category are included Varanasi (earlier Banaras) in eastern UP, though conditions are now changing fast, Muzaffarpur in north Bihar and Puri in coastal Orissa. As for the third category it could be sub-divided into three broad types of regions: rainfed paddy areas such as Deoghar district in the hills of south Bihar and Keonjhar in the interior of Orissa; Satara district in Maharashtra as an instance of a semi-arid zone and finally, Jodhpur, a case of an arid zone in western Rajasthan.

This sample covers a fair number of the main types of problems encountered in rural India in terms of historical background, physical conditions, crop patterns and techniques and social factors, particularly castes, population density etc.

After *space,* comes *time.* I began living or travelling in India in 1952. For several years I confined myself to rather broad studies until 1963, when I started these village and district surveys. Uttar Pradesh, Bihar, Andhra and Tamil Nadu were taken in a first round in 1963-64 and 1967. During a second round in 1978-79 I added some new areas in Assam, Orissa and Rajasthan. Then in 1985-86 (October to February) in addition to another survey of all areas (except in Assam), I added studies of rainfed paddy districts in hilly terrain. In between these three main rounds, I visited Bulandshahr district on several occasions, in 1975 Satara and Varanasi, and in 1982, Thanjavur.[7]

[7] For the first survey see my *Studies in Indian Agriculture: The Art of the Possible* (Berkeley, University of California Press, 1968 and Bombay, Oxford University Press, 1968). For the second survey see *India's Changing Rural Scene* (New Delhi, Oxford University Press, 1982).

These re-surveys help considerably in assessing what is really happening or not happening in a village. Often I met the same people and observed changes in their living conditions. Besides, the more accustomed one becomes to such revisits, the easier it is to grasp, certainly not all the aspects of village life, but at least the main issues with which villagers are concerned.

As far as lodging is concerned, whenever we go to Mirpur (Bulandshahr), my wife and I stay in the village as we have sometimes done in other villages (Varanasi, Satara). Otherwise we comute from the district town. In Hindi-speaking areas, I had no problem of communication and could deal directly with the villagers. In other areas I had to rely upon an interpreter, often an official from the district headquarters.

The three major surveys were carried out while travelling in my own car. These long hours of driving from one place to the other enabled us to have at least a brief look at so many other parts of the country which is not only of interest but helps one to make the link from one area to the other. What contrasts there are, for instance, between those endless and barren plateaus of Karnataka or Maharashtra and the green delta of the Cauvery in Tamil Nadu. The sight of small towns is no less revealing: the booming bazars of Punjab and coastal Andhra with so much trade, workshops, small industries; the quiet small towns in the hills of Orissa or the over-crowded district towns of the plains in north Bihar where small industries are not so conspicuous.

This type of approach also helps one to understand the magnitude of and the fantastic complexity in running such a huge and diverse country as India. In 1985-86, after more than four months in the field and 14,000 km. of driving, it was a strange feeling to be back in Delhi, discussing what I had seen and the broad policy issues in the offices of Krishi Bhavan or in the Prime Minister's office.

What is the value of this approach? First I must confess a loneliness which I have experienced for many years. In August 1985, I attended a very interesting conference in Bangalore on macro and micro rural studies. Between macro studies and villages

[8] Conference on 'The Measurement of Rural Economic Change: Differences in Approach and Results of Village Studies and Large-Scale Surveys,' Bangalore, August 1985, organised by the Social Science Research Council, New York and the Indian Institute of Statistics.

studies, I did not know where to sit![8] Yet I feel that it would be worthwhile for others to follow a rather similar path. This of course does not exclude other approaches, but why not try?

Now, I am well aware of a number of weaknesses in such surveys and re-surveys. The anthropologist may regret that they are not conducted in greater depth, the economist that there are not enough data.

But, such a method takes us away from classical academic work or research *in camera,* to bring us into close contact with the life of the people, peasants or officials, and with the perception of their problems involving so many factors: traditional values, social tensions, human relations, prices, technical questions, etc. Such surveys help in bridging the gap between urban elites and rural people. In other words, one becomes more practical, less prone to a priori ideas, so common not only to academic circles, but sometimes — though fortunately more rarely now — among policy makers. And to conclude, I must confess that I simply love this type of life—away from the routine and rather dull ways of life which one can experience in Europe!

As will be evident in all villages, the yield of crops per hectare are one of the main variables of growth. They are also one of the main indicators of the level of growth already reached. For the reader not too familiar with yields, I would like to give here some broad reference data of yields obtained by farmers on a rather large scale.

1. *Traditional techniques only*[9] (kg. per hectare [ha.])

Good yield, meaning a relatively adequate water management (often irrigation), careful tillage, weeding, some farm manure.

Paddy	2,000 – 2,500	paddy/clean rice 3:2	
Wheat	1,200 – 1,500	wheat/flour	100:85
Maize	1,500		

Medium yield, usually no irrigation, mostly rainfed crops.

Paddy	1,500
Wheat	1,000
Maize	1,000
Millet	700

Low yield, lack of water and/or lack of drainage for paddy, possibly also, weak agricultural practices; usually rainfed crops only, with uncertain rains.

Paddy	500 – 1,000
Wheat	300 – 700
Millet	200 – 500

[9] These yields refer to various parts of Asia before the advent of technical innovations.

2. *Semi-modern agriculture in Asia*, relying on new seeds, adequate doses of chemical fertilisers, if necessary, pesticides, good irrigation, often with pumps; not necessarily tractors; several traditional practices maintained for sowing, weeding, harvesting.

Good yields

Paddy	5,000 – 6,000
Wheat	4,000 – 5,000
Maize*	4,000 – 5,000
Millet*	4,000 – 5,000

*Hybrid seeds to be renewed every year.

Medium yields, water management not always adequate, use of chemical fertilisers below recommended doses, and often unbalanced use of N and P.

Paddy	3,000 – 4,000
Wheat	2,000 – 3,000
Maize	2,000 – 3,000
Millet	2,000 – 3,000
Cash crops	
Groundnut	600 – 1,000 unshelled
	shelled = 2-3; oil content 40-50 per cent
Cotton lint	50 – 100 low yield
	300 medium
	500 good
Sugarcane	30,000 (sugar content 10 to 15 percent)
	50,000 fair
	80,000 – 100,000 good

References to Western countries and Japan

Wheat	: USA average 2,300
	West Europe 4,000–6,000
Paddy	: Japan 5,300 – 6,000
Hybrid maize	: USA, West Europe 5,000-7,000

Chapter Two

'A Village of Ordinary Appearance' in Western Uttar Pradesh

How have they managed, the peasants of Mirpur, 'an average size village of ordinary appearance'[1] lying in the vast plain stretching between the Yamuna and the Ganges? In 1860, when surveying Bulandshahr district, Robert Currie, Settlement Officer, counted a population of 451 on a total area of 276 ha. Today, on the same area, live 1,848 people who, on the whole, are better off than their forefathers.

Mirpur illustrates three major themes of this book. First, the timing of growth and changes. Second, the relation between population increase and technical innovations. Third, the magnitude of the global process of rural development, relying on a rather fast growth for the major crops and on a substantial expansion of other activities leading to a greater diversification of the economy.

The area relies mostly on wheat as a winter crop (*rabi*), on maize, *jowar* (sorghum) and some *bajra* (pearl millet) during the monsoon (*kharif*). Sugarcane, the main cash crop, is a one year crop. Other crops are mustard and gram (winter crops), and pulses. There has always been some double-cropping, a trend which expanded considerably, so that by the 1960s most of the maize, *jowar* and *bajra* crops were followed by a winter crop.

The land-man ratio

The land records and the census reports enable us to present the following data.

As will be explained later, the quality and efficiency of irrigation has considerably improved with the advent of tubewells in the 1950s and 1960s.

[1] *First Settlement Report*, 1859-60.

	Population	Density per sq.km.	Cultivated Area (net ha.)	Irrigated Area (net ha.)	Inhabited Area (ha.)	Uncultivated Land (ha.)
1861	451	163	228	59	3.5	44.5
1921	731	264	264	131	4.0	8.0
1961	1,227	444	260	192	9.0	7.0
1981	1,848	669	250	250	19.0	7.0

Geographical Area: 276 ha., unchanged, at least since 1890.

In 1860 fallow land and jungle amounted to 36 ha. By the time of the second settlement (1889) all that land had been reclaimed. From that time, uncultivated land referred to roads, two tanks near the village and communal land (1.5 ha.), The latter was granted as homesteads to Harijan families after 1961. All these data contain some margin of error, though small, because some plots classified as homesteads (*abadi*) can be kitchen gardens, and some cropped land may have been further converted into homesteads.

Population pressure has had the usual effect on agricultural holdings, i.e., a growing division of the land:

Number of agricultural holdings

1916	47
1965	133 (After the consolidation of holdings)
1986	179

		Number of holdings	Cultivated area (ha.)
1965 :	0.01 – 1.50 (ha.)	70	50
	1.50 – 2.00	11	20
	2 and above	52	180
		133	250*
1986 :	0.01 – 0.5 (ha.)	40	11
	0.5 – 1	50	36
	1 – 2	52	71
	2 – 4	32	93
	4 – 8	3	13
	8 and above	2	20
		179	244*

*The difference between two figures is due to sales of land to non-residents and to new houses. The exact break up of holdings should include land owned by Mirpur farmers in nearby villages. Yet, this would not alter the overall picture. Land owned outside may be around 10-15 ha.

The abolition of the zamindari system implemented in UP in 1952 had hardly any impact on land distribution. The three zamindars kept the land they were cultivating amounting to a total of 36 ha., which was below the ceiling. As for other farmers, they retained the land upon which they had a right of occupancy. From being tenants, they became full owners.[2]

Obviously there has not been a polarisation of the land between the rich and the poor, but a growing process of division of holdings following the death of the owner. For instance, after the death of Tez Pal Singh, one of the main ex-zamindars in the 1970s, his holding, one of the two largest in Mirpur, was divided by his three sons. Each of them inherited 3.75 ha. The same process goes on in the lower brackets.

Even apart from the lack of strong political currents favouring land distribution, the sheer division of holdings makes land reform less and less possible unless one were to adopt a ceiling around 2 ha. (In UP, for irrigated land it had been reduced to 6.3 ha. in the 1970s.)

As far as caste ownership is concerned, one does not come across striking changes. The Jats were and remain the dominant caste. They are the largest group of the population (about one-third of the total population) and own the larger part of the land, although it has fallen from 150 ha. in 1961 to 138 ha. in 1979. The Lodha Rajputs have increased their share from 28.5 ha. to 35.25 ha. The share of the Brahmins has decreased from 16.25 ha. to 13.0 ha.; that of the Jativs (Scheduled Caste) from 20.5 to 19.

Today, as in the past, sharecropping and other tenancy arrangements play but a minor role. Two of the sons[3] of Tez Pal Singh (ex-zamindar), after inheriting land following the death of their father entered sharecropping arrangements with some farmers living in a nearby village. Smaller holdings in some places also come under sharecropping when the owner is old and his sons are away.

On the whole, Mirpur, like other villages around, depends mostly on cultivating owners. As regards landless people, there were about

[2]They could become full owners by paying the state eight times the rent they were giving to the zamindar, or become tenants of the state. Later on in the 1970s the difference between *bhumidars* (proprietors) and *sirdars* (tenants of the state) was abolished. With the abolition of zamindari came the abolition of various feudal rights, such as *begar,* unpaid temporary work for the zamindar.

[3]They live and work in the town with their families.

seventy out of about 200 families in the early 1960s. They included artisans like potters, carpenters, Naīṣ (barbers) and Scheduled Castes : Dhobis (washermen); Bhangis (scavengers, sweepers); Jativs, formerly called Chamars, being mostly agricultural labourers. The latter were and remain the largest group belonging to the Scheduled Castes.

The total population of Scheduled Castes has increased from 308 in 1961 to 531 in 1981. The rate of increase seems a little higher than for non-Scheduled Castes, but one should bear in mind that among the people who have left the village, the proportion of Scheduled Castes is lower than the other castes. This is confirmed by the substantially lower rate of survival of children of the poorer people, who so often are Harijans. In the early 1960s one would come across Jat or Brahmin couples with only a few children dead compared to four or six alive, whereas it was often just the reverse with Harijan couples, usually the landless. Today, while the gap may have narrowed slightly, differences remain substantial.

An early start

Although Mirpur is nearly entirely outside the command of the irrigation canals built between the Ganges and the Yamuna in the 19th century, the economy did not remain static. It was indirectly affected by the overall growth which occurred in Bulandshahr and in the other districts of western UP: the progress of sugarcane and wheat thanks to irrigation, expansion of bazars and cities and better means of transport.

Though moderate for several decades, the population increase led to the reclamation of all cultivable land by 1889. Such a trend is found practically everywhere in the world. What does not follow automatically is a change of techniques stimulated by population growth as claimed by Esther Boserup,[4] a topic to which we will return at the end of this book. However, in Mirpur, as in most of north-west India, technical changes did occur within traditional agriculture, following a pattern which can be found in many countries in ancient times.

[4] E. Boserup, *The Conditions of Agricultural Growth* (London, Allen and Unwin, 1965).

Irrigation techniques improved in two ways. Masonry wells gradually replaced temporary, often choked wells which delivered less water. Around 1930, the *mot* system (leather bucket pulled by a pair of bullocks) was replaced by the Persian wheel, which is more efficient. The cropping pattern was altered, cotton going down in favour of the expansion of sugarcane, thus yielding a better return. Later, in the 1930s, new cane varieties originating from Java and Sumatra, were multiplied in India and replaced the old varieties in the north-west. Improved varieties of wheat followed.

As far as agricultural implements are concerned, few changes occurred during that period of time: the wooden plough strengthened by a piece of iron, the plank for levelling, the seed drill, remained as they had been for centuries. The equipment (*kolu*) for extracting sugarcane juice and transforming it into raw sugar (*gur*) was modernised through the use of the iron crusher (instead of stones) rotated by bullocks and the use of a large iron pan placed on a fire of sugarcane waste (*bagasse*).[5] To cut fodder, farmers used a chaff-cutter worked by hand.

Farmers used part of their cow-dung for cooking and part as farm manure (badly kept and exposed to the sun).[6] It was spread once in the year during the monsoon or before planting sugarcane.

Until the 1950s technical innovations were typical of the changes which have occurred in many parts of the world over thousands of years. These 'small changes'[7] though not negligible, may provoke increases in yields of 10-30 per cent, rarely more.[8] It shows how simplistic, if not wrong, is the idea of an unchanging or static rural world in ancient history.

The scene began to change around 1950. The main ex-zamindar

[5] I do not know when these changes were introduced, probably towards the end of the last century or a little later.

[6] Manure pits are not well arranged. Hence, there is much loss, a defect rather surprising among such good farmers but one that is widespread even now in most parts of India.

[7] This phrase is used by the French historian Guy Bois when studying a village near Cluny, in France, from the 10th to the 18th century. See P. Gourou and Gilbert Etienne, eds., *Des labours de Clunny à la révolution verte* (Paris, 1985).

[8] The shift from the *mot* to the Persian wheel is more productive; the former can irrigate 0.10 ha. a day, the latter 0.25 a.

installed a tubewell run with an oil engine, an example later followed by two other large landowners. Then came a state tubewell. The first development block in the district was created in 1955 in the area of Mirpur, with its headquarters at Unchagaon, a big village 3 km. away. Chemical fertilisers began to be introduced on a limited scale.

By 1955, a metal all-weather road was completed between Unchagaon and the nearest small city, Jahangirabad, 12 km. away. The isolation of the area, especially during the monsoon, was broken and this had some impact on development.

In 1963-64, during our first survey, yields of wheat, the main cereal, varied between 740 and 2,300 kg./ha. The former referred to unirrigated wheat relying on the short winter rains, the latter to yields from irrigated fields with small doses of chemical fertilisers. The most common yields averaged 1,200-1,300 kg./ha.; yields of maize were a little less. Sugarcane yielded around 6,000 kg./ha. (in terms of raw sugar, i.e, one-tenth of the cane weight) for the first crop and about 3,600 for the second or ratoon crop. Such yields are typical of a good traditional system with only very few technical innovations. It was very clear at the time that to increase yields further, in view of the accelerating population pressure, more radical changes would be needed than the 'small changes' of the past.

The supply of irrigation was not adequate. In the dry season, when wheat and sugarcane need water at the same time, the delivery system of water would be too slow especially for holdings of above 3 or 4 ha. In order to introduce new seeds and chemical fertilisers, one would require more water than for traditional agriculture. Thus, more powerful means were needed, such as tubewells, a point clearly perceived by many farmers. It was no less clear that with the fast growing population the whole area had to be irrigated along with a higher consumption of chemical fertilisers.[9]

On the whole, the peasants of Mirpur were ready to enter a new phase of their history. They were starting from a fairly high level, relying on solid and efficient agricultural traditions. The Jats (the dominant caste), had been well known for such qualities but, most

[9] Average rainfall: 660 mm. in forty days. The bulk of the rainfall is during the monsoon from July to September. There are scattered and unreliable winter rains from December to February.

other castes, Brahmins, Vaishyas, Lodha Rajputs, were equally hard working. To sum up, there was the ideal combination of growth factors : good alluvial soils, a large potential for irrigation, a fairly advanced stage of development and excellent agricultural castes. However, we find here an illustration of T.W. Schultz's book : the traditional techniques had practically reached their ceiling. To proceed further, technical innovations were badly needed.[10]

The new growth process

Should one talk of a 'breakthrough' or a 'green revolution' as is so often done? Both terms are not very adequate. They convey an idea of abrupt socio-economic change in production and in life styles. In Mirpur, these words are both correct and not correct. It is true that in a few years time, the yield of wheat doubled, a breakthrough in itself, but, as will be seen, other socio-economic changes have been gradual, more properly described as an evolution than as radical changes.

The first change was the consolidation of holdings, which was being gradually introduced in the districts of UP. The operations were being conducted during our first stay in Mirpur in 1963-64. It is an extremely tedious and complex job, involving litigations or, better, the settlement of litigations which had often been dragging on for years, or which came up at the time of the consolidation. The attribution of the new plots is no less difficult. Finally, it is no mean business to consolidate, as in Mirpur, 613 plots (*chaq*) into about 135. Complex in themselves, these various tasks may give way to all sorts of abuses and to corruption.[11]

The first and immediate result of the consolidation of holdings completed in 1965 was to lead to the construction of many private tubewells, an investment being less attractive when the owner had several scattered fields.

By a lucky coincidence the first high yield varieties of wheat (from Mexico) appeared just after. Karan Lal, one of the local

[10] T.W. Schultz, *Transforming Traditional Agriculture* (New Haven, Yale University Press, 1964). He points out that any system finally reaches its limits, hence the need for innovations.

[11] In Mirpur, I did not get complaints. The assistant consolidation officer conducted the operations honestly.

netas with close contacts with the administration, did a first trial with the *rabi* in 1966-67, with a few others. They got such a superb crop that by the following year, most of the fields were sown with the new seeds. At the same time, the consumption of chemical fertilisers rose considerably. Thus, the three interdependent key elements—water-seeds-chemical fertilisers—began to spread. This point must be emphasised. Traditional and even improved varieties of wheat[12] have a rather limited response to chemical fertilisers[13] unlike the new seeds. But, at the same time, if one uses more chemical fertilisers, one needs more water than for traditional grain.[14]

Another important factor is electricity. By 1963, the electric network was spreading but it was confined to tubewells. Only later did the villages proper receive electricity. Power supply was a great help because electrical tubewells are cheaper than tubewells using oil.

The process of growth kept on increasing. Already, in the 1970s, the entire surface had been covered with private tubewells ensuring better crops and more double-cropping, some of the tubewells being installed with the help of cooperative loans. In 1978-79 there were forty-eight private tubewells and in 1985-86, sixty-three; several of them belonging to owners with holdings below 2 ha. Most of them were 5 HP and had a command area of 5 ha. The depth of the tubewell varied between 15 and 50 metres. Part of the recent increase is due to the shortage of electricity referred to later.

Consumption of chemical fertilisers and new seeds of wheat, reached all the farmers including very small landowners of 0.5 hectares or less, although the doses of fertilisers applied by small farmers could be below the recommendations, for lack of money.[15]

[12] Not to confuse this with the new high yield varieties.

[13] The better response of new seeds is due to genetic factors and due to the fact that they are dwarf which reduces the risk of lodging.

[14] 5-6 waterings instead of 3-4.

[15] On the whole, there are few pest attacks on wheat which grows in the dry season. The main problem is to renew the seeds every four to five years in order to avoid seed deterioration leading to diseases like rust.

Average wheat yields per hectare

1963-64	1,200 – 1,300
1975-76	2,000
1978-79	2,500
1985-86	3,000 – 3,200

In 1986, some farmers were able to harvest only 2,000 kg./ha. due to unsufficient irrigation, inadequate application of chemical fertilisers, or light soils. Most farmers would harvest between 2,560 to 3,200 kg./ha., but some of them would reach 3,800.

Progress in maize has been much less spectacular. The new hybrid varieties tried in the late 1960s did not prove well suited to local conditions, a situation not altered until now. As a result, farmers resorted to their traditional varieties.[16] Thanks to better irrigation and small doses of chemical fertilisers, yields have increased from about 1,200 kg./ha. to 1,500 or more often 2,000 kg./ha.

Jowar and *bajra* are increasingly cultivated for fodder and harvested green.[17] Sugarcane, for long the main cash crop, has not benefited from the breeding of new varieties giving a higher yield. In addition, the climate of western UP with its cold spell in December-January[18] is not favourable to sugarcane, which reduces the juice extraction rate. Irrigation and chemical fertilisers have helped increase yield, although moderately : 6,500-7,000 kg./ha. for the first crop (raw sugar). However, in the past two years severe attacks of insects (pyrilla) have brought a sharp drop in yields all over western UP : down to 3,000-4,000 kg./ha.

The area devoted to gram and mustard has fallen in favour of wheat, though the latter is now picking up. There have been no new and better seeds for the former. As for mustard, the recently released varieties provide a yield increase of 20 per cent over traditional varieties, which leads to less profit than in the case of a sharp increase as in wheat. In addition, the new variety seems more vulnerable to pests, complain some farmers.

Some short duration (sixty-five days), new varieties of pulses (*mung* beans) can be squeezed between the harvest of wheat in

[16] Maize is vulnerable to heavy rains, especially at the beginning of its growth which, in certain years, reduces the yield.

[17] A trend accompanying the rising number of buffaloes. See later.

[18] Sugarcane—raw sugar (gur) ratio, 10 : 1.

April and the advent of the monsoon with the *kharif* season in July. Yet, success has so far been limited, because of lack of time,[19] especially when threshing machines for wheat cannot be used due to power shortage (see later). As a result, farmers have to revert to the time consuming process of threshing with the help of cattle. *Mung* beans may also be attacked by pests.

In Mirpur as in so many villages, farmers lack the tradition of a kitchen garden,[20] thus wasting small plots of land close to their house. The situation has begun to change and in 1985-86 one came across vegetables cultivated by several farmers. They irrigate them with the hand pump, usually installed next to the house for drinking purposes. It is also interesting to come across many new trees (eucalyptus) planted in the village, outside or along the field bunds.[21] There have always been some rather poorly maintained mango orchards. Now, near the houses there are more papaya trees.

Potato appeared in the late 1970s as a consequence of a growing demand and better transport facilities. It is cultivated between October and December, after the harvest of *jowar* and before the sowing of wheat. This postpones wheat sowing from November to late December. Although wheat yields are reduced by late sowing, this is more than compensated by the sales of potato. Four cold storage houses have been built in Jahangirabad, 12 km. away from Unchagaon. Such a cash crop (12-16 t./ha., a yield which is moderate, but not so bad for a new crop) brings more income to farmers, but they complain of sharp price variations. The glut in the market in 1985 brought down the prices.

In the past twenty years animal husbandry has known considerable growth. Buffaloes are more numerous and give 4-6 l. of milk per day, or more in the case of better bred and better fed animals. A few more cows, cross-bred with Frison, have appeared and are admired for their very high yield : upto 18 l. per day while local cows (rare in Mirpur), often give not more than 1-2 l. Sales of milk are expanding. Several men bring it (50-70 l.) on their bicycles to

[19] *Mung* beans require much labour.

[20] Kitchen gardens are more prominent in West Bengal and in certain districts in the north-east of the peninsula. See Chapter 10.

[21] This is a result of several government re-afforestation schemes. At a subsidised rate, one sapling is sold at Rs. 0.10 (actual cost Rs. 0.25). After five to six years, the tree is sold for Rs. 200.

Unchagaon from where chilling units collect the milk and send it by truck to dairies in Bulandshahr or Delhi where the milk is pasteurised. When we were in Mirpur a private dealer had just opened a shop to collect milk in the village and dispatch it further.

In addition to old sources of fodder—the top of the sugarcane plant, *jowar, bajra,* maize stems and leaves, all cut in the chaff-cutter —farmers are now able, thanks to increase in yields, to somewhat reduce their wheat area in favour of small plots of *berseem.* Between October and April, with the help of small doses of chemical fertilisers, farmers have five to six cuttings of *berseem.*

While bullocks are no longer used for irrigation with the disappearance of the Persian wheel, they still play some role in ploughing. Four farmers now have a tractor each which they use and rent to others. Between half and three-quarters of the cultivated area is now ploughed with tractors. It may happen that the first ploughing is done by a tractor, and subsequently with bullocks. The latter are also increasingly replaced by buffaloes for pulling carts. The old cart is giving way to the *bogey,* a cart with two wheels with tyres. One buffalo has replaced a pair of bullocks and even though it is slower, the *bogey* is more profitable and easier to pull. As alluded to above, threshing is now less and less dependent on bullocks, the latter being replaced by small machines manufactured in Bulandshahr and other towns. This is a good example of the simple technology of small factories and of the impact of rural electrification.[22]

From a technical point of view, what is striking—and now very common in all advanced areas of Asia—is the very sensible mix of old and new techniques. Farmers have started with the most decisive innovations in order to increase production, i.e., tubewells, new seeds of wheat, chemical fertilisers. Mechanisation has, in recent years, expanded with regard to land tillage and threshing. However, many ancient practices still remain in use : some amount of ploughing with bullocks, sowing with the seed drill attached to the wooden plough, weeding by hand, harvesting with

[22] Small is nevertheless not always beautiful! These machines lack a proper safety device. As a result there have been cases (one in Mirpur) of workers having one hand cut in operating the thresher. These machines reduce waste by making it possible to store the grain more quickly. See, J.A. Allan, Appropriate Technology and Labour Supply in Rural India, *Journal of Tropical Agriculture,* University of Singapore, December 1976.

the sickle. Sugarcane is often still cut by hand. Wheat, cane and maize are often carried in bundles on the head or on *bogies.*

Growth and diversification outside agriculture

The process of growth and diversification observed within agriculture, including cattle and milk, is linked to similar trends outside agriculture.

The expansion of modern traffic (trucks, buses, tractors), is striking on the district road from Bulandshahr to Anupshahr. At the crossroad where one turns towards Jahangirabad, except for the usual gathering of monkeys, the road is hardly recognisable compared to our first arrival in 1963. The metal road has been widened and asphalted as far as Unchagaon. The bypass avoiding Jahangirabad, constructed in the late 1960s, is now heavily overcrowded. Countles garages, mechanical workshops, factories and tea stalls, are mushrooming on both sides of the road.

Further, towards Unchagaon, twenty years ago one would have met practically no motor vehicle, except here and there a government jeep. The bullock cart dominated the road and it was not easy to overtake them, so frightened were the bullocks! Today, the traffic is a mix of trucks, *tangas, bogies* and bullock carts. The animals have become used to motors, and mingle with them quite placidly!

Whenever we come to Unchagaon, the bazar street has grown a little more towards the canal. In 1963, if we wanted to buy good fruit and vegetables we had to go to Bulandshahr. Today, we find all we want in Unchagaon. All sorts of shops have appeared; tailors, merchants of cloth and ready-made shirts, shoe makers, mechanics and electricians. There are more carpenters and saw mills. Several people sell and repair bicycles. There are several barber saloons, along with a photographer and radio dealers. The number of chilling units where milk is collected and sent by truck to Bulandshahr or Delhi keep on increasing. The number of tea stalls, of sweet shops and general merchants is also rising. The Punjab National Bank, has also opened a branch here.

Deeper into the countryside, new roads made of brick are being built since our survey of 1978-79, so that fair-weather, dusty or muddy roads are shrinking. In the next few years they should have practically disappeared in the whole region. Now this is no mean progress, because until recently, the 3 kilometres between Mirpur and Unchagaon were so bad that a buggy could carry only half its normal capacity of 16 quintals. Between Unchagaon and Mirpur, trucks remain very rare and there is no regular bus service yet. Carts, *tangas* and bicycles which are increasing remain the main means of transport.

The whole process—trade, small industry, transport—has been helped by the advent of electricity in the 1960s. Today all villages are covered which paves the way for several kinds of workshops and for better irrigation.

The age old practice of grinding grain every morning for a few hours through a mill stone, a task done by the women, is disappearing. There are now six small flour mills in Mirpur. Two belong to Lodha Rajputs, rather small land-owners, the others to poor people: a Muslim, a Jativ (under the IRDP loan-cum-subsidy)[23] and two others.

Bullocks have practically disappeared from the *kolus*. Cane crushing is now done with a motor, usually one attached to a tubewell. There are four of them, much less than in the old system. Some farmers increase their income by crushing cane for others and then transforming the juice into raw sugar (*gur*), following the ancient practice. One must add that more farmers now deliver their cane to new sugar mills built in Jahangirabad.

Part of the old rotating equipment of the Persian wheel is now used with bullocks to run the chaff-cutter (previously operated by hand) to cut fodder for animals. As mentioned earlier, most threshing, when power is available, is done with small electric machines. A young Muslim has opened a workshop for repairing the pumps and motors.

If electricity has been an important factor of growth for the whole rural economy, it has rapidly become a mixed blessing. At least since the mid 1970s, constant complaints are heard about breakdowns, falling voltage, poor maintenance and repair of transformers, of electric lines. In addition to such shortcomings, the supply cannot cope with the demand, a situation which worsens further in the event of a poor monsoon.[24] This problem

[23] Integrated Rural Development Programme. Special schemes for the poor, see Chapter 3 and 4.

[24] The area relies on hydro-electric power.

comes up practically everywhere in India. Many shortcomings could be corrected by more efficient and more determined state electricity boards, but farmers also have their share of responsibility considering the amount of pilfering. The meters were tampered with so much that finally the UP government adopted a flat rate system. Now the farmers grumble because they have to pay regardless of whether power is available or not.[25]

In all fairness, one must recognise that to operate this enormous network through so many villages, so many connections to homes, to tubewells (in Mirpur sixty-three spread over 2.76 sq. km.) is a tremendous job, especially when considering how fast rural electrification has grown which, all over India, started from naught in the 1950s. But the speed with which electricity has expanded is precisely one of the causes of the weaknesses of the system.

When the first troubles started, power shortages could last for weeks and, when occurring in the hottest months of May and June, one can imagine how sugarcane suffered. Now the UP electricity board is trying to *organise* the shortage by supplying power eight hours out of twenty-four during the day for one week and eight hours out of twenty-four during the night, the next week. The latter period is obviously inconvenient for the users, but they have to cope with it.

Falls in voltage are no less troublesome. If the engine is not switched off immediately it can be badly damaged, as often happens when the farmer is away in the fields. As a remedy many tubewells have been equipped with a thermal protection device which automatically disconnects the engine in case of lower voltage, an extra cost for the farmers. The limited supply of electricity induces farmers who can afford to either install a second tubewell or buy an oil engine to replace the electric one temporarily. All these purchases mean over-investments. Today, the total command area of sixty-three tubewells is equivalent to 315 ha. (5 ha. per unit) whereas the actual cultivated area is around 250 ha.

To conclude, after stimulating rural development considerably, electricty has become one of the major factors hindering further development. As will be seen, this is a universal problem all over

[25] The state electricity board replies that the rate takes into account the limited supply of electricity.

India, although its magnitude varies from region to region. Several areas are much worse off, as in Bihar, others manage better, as in Punjab.[26]

Could one pin hopes on alternative sources of energy? In 1978 one farmer set up a biogas installation, supplying gas for cooking and light in the house. It did not operate too well in winter because of the cold (at night the temperature falls to a few degrees above zero). One or two more installations followed later on, but again in 1985-86 I heard complaints. The future of biogas in the area does not look very promising.

Until the 1960s only the fairly well off farmers owning 5-6 ha. or more could live in a brick house. All the others lived in mud houses. Part of the money earned through new wheat has been invested in construction. Today most farmers owning around 1-2 ha. of land live in brick houses. Brick making has expanded through small units in several villages. Many landless agricultural labourers have found additional employment in brick making and home construction.

Trade expansion is percolating down from Unchagaon to Mirpur. In 1963, there were only a few tiny shops selling salt, matches and some grain. In 1978, half-a-dozen small shops had been opened. In 1985-86 there were eleven of them, selling grain, chilli, onion, soap, cigarettes, batteries, biscuits, candles, even some cloth. It is refreshing to note that four of them belong to Harijans. Several other shopkeepers are also either landless or small landowners. Only one of the shops belongs to a medium farmer, a Brahmin. Pedlars have started visiting villages, their bicycles heavily loaded with cloth or fruit.

So far, traditional means of transportation are still expanding. In addition to landowners using their own *bogey*, eight people have, in the past ten years, acquired a *tanga* and a horse, most of them thanks to an IRDP loan-cum-subsidy. They all belong to the weaker sections of society. They ply as far as Jahangirabad and back with a load of goods like *gur* or with several passengers.

Job opportunities outside Mirpur

For a while now some villagers have been finding a job outside : in

[26] This problem is by no means confined to India. One finds it in other Asian countries like China and, more recently, Pakistan.

government service, in business, in the army.[27] Such a trend has gained momentum with the development occurring in the district and further, as far as Delhi, which has grown so much since 1947.

Number of men working ouside		*Total population*	
1963	*47*	*1961*	*1,227*
1986	*142*	*1981*	*1,848*

Obviously the rate of emigration has accelerated, bringing more income to Mirpur. In 1986, seventy-nine men were working outside and had taken their wives and children with them. Sixty-three men had left their families at home with their parents or brothers, sending money home to help them. Some people have found employment in industries in Delhi or other big towns. Several of them have become clerks and other government servants. Others are in the army or the police. Many of them have found jobs as teachers in Unchagaon or in neighbouring villages. Several people have opened shops in Unchagaon. The son of a traditional barber (*naī*) has opened a hair dressing saloon in this locality.

From our data, higher castes seem to find a job outside more easily, perhaps due to their better education, and connections as compared to the Scheduled Castes : 8 men belonging to the Scheduled Castes out of the 79 who left with their families, 11 out of the 63 who left without their families. As will be seen in all villages it can considerably help the family of a small and even of a medium farmer if, at least one of the sons has found a job outside and sends money home. This is even more important for the landless families.

[27] Bulandshahr in an important recruitment district for the famous Jat Regi ment.

Chapter Three

The Impact of Growth: The Poor, the Medium and Rich Farmers

What has been the impact of so many economic changes on the people? How have the living conditions of the landless labourers, of the small and rich farmers, the Scheduled Castes and the other castes been affected? Before taking individual cases, let us start with some broad observations based on meetings with groups of people.

From the early 1970s, we could notice changes in conditions of living which gradually became stronger and wider, but did not follow a single pattern. For instance, a number of poor people have become less poor, while some others, even today, struggle for survival under very difficult conditions. Most medium farmers are clearly better off, but some of them are under strain due to the falling size of their holdings.

A first change concerns caste relations. Basic distinctions, particularly for marriages, remain as they have been for centuries. Marriages are arranged within the same caste with a boy or a girl of another village. With regard to food, the rules are relaxing. Scheduled Caste people may eat with others in social functions, (*dawat*). In social gatherings (1963-64) Jativs, Dhobis and other Scheduled Caste members would squat next to the *charpais* (traditional bed) where the Brahmins, the Jats and the Lodha Rajputs would sit together. Today it is no more so. Harijans sit on the same *charpais* with anybody.

In 1963, when questioning Natu, a tailor, and one of the leading Jativs, I was told: 'the difference between now and the past is that we no more have a lock (*tala*) on the mouth.' Today Jativs and others have become not only outspoken but also more conscious of their rights, less subdued. Several of them have improved their conditions of living, being better dressed and fed. More of their children go to school, sometimes to high school and even college when they get a scholarship. In the Janata Inter-College in Mirpur which covers several villages, there are now 478 students of which seventy-nine belong to Scheduled Castes.

The Bhangis (scavengers, sweepers) are less affected by the process of development. Several of them remain extremely poor: shabbily dressed, living in badly maintained mud houses.[1]

Among upper caste people, some are now cleaning their homes themselves, which deprives the Bhangis of their traditional job.[2] The Bhangis are also less educated than other Scheduled Castes. They seem to have more difficulty than the Dhobis in finding jobs outside, especially in services, a result of their poor education.

These issues deserve detailed studies. From my observations I cannot do more than hint at the possibilities of certain castes, taken as a whole, doing better than others in deriving benefits from economic changes.

In discussions with Jativs and others there came up again in 1985-86 as in my previous visits, a ticklish problem. They complain that they are not completely employed in spite of some progress, while landowners keep grumbling that; 'these people are lazy, they do not want to work.' There may be cases where wages offered are considered too low—here again, in spite of an upward trend in real terms—but we cannot also exclude cases of possible inertia.

Another issue, and this one is new, is the fact that the demand for agricultural labour, after rising in the early years of the new wheat, has now begun to fall due to the rising number of small holdings where non-family labour may not be needed except for the wheat harvest. Such a trend, if any, should become clearer after some more years. It shows in any case how the employment situation remains difficult even in a progressive area.

When summarising wages and prices we obtain the following data:

Wages

	1963-64	1978-79	1985-86
Daily agricultural labour (Rs.)	1.0	5.00 to 6.00	10.00-12.00
in kind (kg. of grain)	5.7	10.00	10.00-12.00
Permanent agricultural labour (Rs. per month)	30.00	100	220.00-230.00
Daily wages for cane crushing and raw sugar making (Rs.)	–	6.00-7.00	10.00-20.00
Daily wages for masons (Rs.)	–	10.00	20.00

[1] They are particularly vulnerable to calamities, as, when during a heavy rain, their houses collapse.

[2] Normally each of them spends part of the time working for a number of families and gets some grain as payment (*jajmani* system).

At the time of the harvest of cereals, labour is paid in kind. More often daily wages do not include food. There may be cases where they fall below the more common rate mentioned above. Wages of permanent labourers may include some additional benefits: some food, clothes, medical care, but permanent labourers are few compared to casual labourers.

Prices

	1963-64	1978-79	1985-86
Retail price of wheat may vary by 5-15 per cent compared to wholesale:			
Selling price Rs./q.	40.00-55.00	110.00-130.00	150.00-200.00
Selling price of milk at Unchagaon Rs./l.	–	1.50	2.50
Gur (raw sugar) selling price ** Rs./q.	80.00	75.00	280.00-320.00

** The price of *gur* was exceptionally low in 1978-79 due to a glut in the market (1977-78 : 100). The price in 1985-86 was exceptionally high due to a steep fall in sugarcane output following very severe pest attacks.

Prices and wages cannot be measured with a great deal of precision due to so many variables in time and work. However, the above data fit reasonably well with national data. The general price index for agricultural labourers increased by 66 per cent between 1978-79 and 1984-85, the food index by 40 per cent.

As will appear more clearly through case studies of particular families, the poor people who have improved their lot have usually managed to increase the number of sources of income, be it in transport, milk sales, construction, petty trade, combined with agricultural labour. It also makes a substantial difference when one family member finds a job outside Mirpur.

Medium and upper farmers

Most landowners with holdings of 1 or 1.5 ha. or more have replaced their mud houses with brick constructions. Clothing keeps improving: women wearing lighter colours, young girls and boys are smart, the latter wearing trousers, synthetic shirts and canvas shoes. These trends, already noticeable in the 1970s, are even more obvious now.

Most of the Jats, Brahmins and Lodha Rajputs falling in that category used (in 1963-64) to eat mostly *chapatis* of wheat or maize, with some spices and *chutney*, some pulses, milk[3] some mustard leaves, *gur* a few mangoes in season. Tea with milk and sugar was a luxury in the 1960s like fruit and vegetables. Today, the former is becoming common, the latter no more so rare. Most people remain vegetarian.

The progress of education is also quite apparent with the growing number of students who have graduated with a Bachelors degree.

As for semi-durable goods, the number of bicycles has increased, transistors and watches have become more common. In some houses one comes across sewing machines. There are now even some television sets.

All these changes are the result of very hard work, cleverness and ability in using all possible means of increasing production and finding additional jobs. A large number of these farmers adopted technical innovations in spite of their illiteracy or very limited schooling.

Amongst the few really rich families are the ex-zamindars. One of them (holding 12 ha. of land) is still alive and leads a very simple life, having spent most of his savings in having his two boys educated in good, but expensive public schools before joining college. Now both of them have found good government jobs in cities.

The other main zamindar (Tez Pal Singh) died a few years ago. His land has been split into three. Two of the sons are outside and have their land cultivated by tenants. One lives in Mirpur. He will have to work hard because he now has only 3.75 ha. Will he be able to keep his motor-cycle, one of the few in the village?

Small, medium and large landowners are trying to make the most efficient use of their land and, until now, they have succeeded in compensating the shrinking of holdings by higher productivity. In 1963-64, a minimum economic holding allowing a family to live reasonably well without additional work was around 1.75 to 2 ha. Farmers owning less than 2.75 ha. usually had no wheat surplus for sales. For cash they relied entirely on sugarcane. Now the minimum economic holding has come down to 1-1.5 ha.

[3] Landless labourers and very small landowners usually drink very little milk except the children. Whenever they can, they sell part of it.

and, out of such a small area, farmers sell not only *gur*, but also some wheat.

Money plays an increasingly important role in the farm economy. Peasants sell more and also spend more cash on inputs: tubewells, electricity, seeds and chemical fertilisers. The government has set up a wide system of subsidies for most inputs and for the purchase of grain. Side by side with the free market, the government buys part of the cereals (wheat and rice) at a procurement price. Sales in government fair price shops found more in cities than in villages, are subsidised.

These measures have played a role as incentives to farmers, but, as will be seen at the end of this book, the whole system is becoming extremely costly, hence the trend to reduce subsidies.

These matters are hotly debated in the evening when Jats and others meet around their *huqqa* (water tobacco pipe). The complaints are increasing compared to my previous visits, in a familiar vein reminiscent of similar complaints on European farms. Yet, when we look at prices, the situation has not deteriorated so much.

Cost of Inputs

	1978-79	1985-86
Urea Rs. /q.*	166	215-225
Di-ammonium phosphate, DAP*	200	335-350
Investment for 5 HP tubewell	7800	11000
Annual maintenance and operation of a tubewell	1050	1500
Electric bill for a tubewell	960	not available
Government procurement price for wheat, Rs./q.	110	162
Free market farm harvest price, Rs./q.	115.50	170
Selling price of *gur* (raw sugar), Rs./q.**	75	280-320
Selling price of sugarcane Rs./t.	70	240-280

*The price of fertilisers is cut by half for cultivators of holdings below 1 ha.
**The first price in 1978-79 was exceptionally low (as compared to Rs.100 in 1976-77) and the latter exceptionally high.

Let us take the cost-benefit ratio on 1 hectare yielding 3,200 kg. of wheat. The total cost including tractor charges amounts to Rs. 2,400. At Rs. 162 /quintal (procurement price from February 1986) the gross profit amounts to 5,184, which means a net profit of Rs. 2,784. If the cultivator uses his own bullocks for ploughing, his cash expenditure is reduced by Rs. 640. Cultivators of holdings below 1 ha. save around Rs. 250 due to the subsidy referred to above.

Sugarcane cost per hectare (with tractor ploughing) comes to Rs. 2,350. If the farmer does not use his own seeds but purchases them, he incurs an extra Rs. 1,000. Unlike the wheat harvest which has to be done quickly (so that most landowners need paid labour), cane cutting spreads over three to four months. It is usually done by the whole family, including women and children, and labour charges are thus limited. With a fairly good crop of 65 t. of cane per ha., the farmer gets a gross profit of Rs. 15,600. A poor crop of 40 t. as in 1985-86 brings Rs. 9,600.

The ratoon crop always has a lower yield (10-15t./ha. less), but expenditure is also less since there is no ploughing.[4]

In sum, the more spectacular increases have occurred in wheat production. Maize and sugarcane are doing moderately better than they were twenty years ago; *jowar* and *bajra* are stagnant. New crops like potato and other vegetables are still playing only a limited role. Milk, on the other hand, has increased substantially, not so much in yield per buffalo, but through the increase in cattle, although I have no precise data on such a development.

Family planning and problems of youth

While in the early 1960s women and men did not have the faintest idea of family planning, the idea has percolated down to the village. But in the past fifteen years actual progress has been slow. True, one comes across cases of men and women having been sterilised. A doctor does the operations at Unchagoan. Some young couples who have moved to Bulandshahr confine themselves to two children, provided at least one is a boy. On the other hand it is disquieting to meet young medium farmers, educated, smart and clever who, in their early thirties have already produced four children or, in one case even six. In this case, the couple had five daughters and one son. The boy may have been born last, a clear case of the impact of traditions. And yet people are well aware of the growing pressure on land and of the troubles their families are facing! They also know how difficult it is to find a job outside Mirpur. In spite of all the progress made so far in the economic field, it is clear that population control is becoming a matter of increasing concern.

[4]To these data one should add the depreciation of tubewells, i.e., Rs. 500 per year, since a tubewell lasts around twenty years.

One of the results of the slow progress of family planning is the emergence of problems amongst the youth. By 1980 itself I heard several farmers beginning to grumble about their sons. Today the problem is becoming serious. The men who made the green revolution are now in their late fifties or sixties. Their hair may be grey but they are still as active and sturdy as they were fifteen years ago. There is my friend Ram Singh, for example, barefoot, wearing an old *kurta* and *dhoti*, working the whole night on his *kolu* because at the moment electricity is available from 8 pm to 6 am. Many other Jat cultivators with holdings of 1.5-3 ha. are like him. Their sons on the other hand, wear smart clothes not so different from the middle urban classes: trousers made of synthetic material, a nice coloured shirt, canvas shoes. Some of them have completed their high school others a B.A. and they sit at home waiting for a white collar job. They do participate in farming, help in cutting cane, push the bullocks which operate the chaff-cutter, but they do not do it with great pleasure and certainly not always. Kacheru, a typical medium Jat farmer, is particularly outspoken, half laughing, half sarcastic: 'These B.A. fellows are useless. They first sit under a fan in their classroom or their hostel. They can no more stand the heat and the sun in the field. They are not interested in tilling the land. What they dream of, is an air conditioned office!'

Later, the local sub-inspector of police raised other complaints. Some of these young people, after remaining idle too long start indulging in thefts.[5] One cannot deny that a number of these boys are just turning soft in comparison to their fathers. Such an evolution is as normal as it is unavoidable, since one finds it in so many societies. As we will see once more at the end of this book, it may contribute to various kinds of political tensions beyond the village world as well. On the other hand, it is premature to pass a definite judgement. The phenomenon is still too recent and we must wait and see what happens when these youth fully replace their fathers as heads of families.

Marriages and indebtedness

Marriages have, for a long time, been a source of indebtedness,

[5] The law and order situation has deteriorated since 1963. At that time we rarely heard of incidents of thefts as one does today. Some dacoits are even resorting to fire-arms.

in view of the heavy expenditure incurred. According to Hindu customs, often also adopted by Muslims in India, the main expenditure is borne by the girl's family.

In the past twenty years the situation has not improved noticeably. It may even have deteriorated in certain cases. When Tez Pal Singh got one of his daughters married, he spent Rs. 15,000. His son who has to manage on one-third his father's land recently spent Rs. 100,000. At the other end of the social ladder, Scheduled Castes now spend Rs. 3,000 to 5,000 for marrying a daughter, compared to Rs. 500-800 during our first visit. Jat landowners with 2-3 hectares of land easily spend Rs. 30,000-50,000 now. Such heavy expenditure[6] brings all sorts of hardship not only to the poor but also to the not so rich, especially when they have several girls.

As for money lending, it varies. In certain cases interest free loans from friends and government loans at 11 per cent interest are used for marriages instead of for productive purposes as required by law. Money-lending often remains at 36 per cent per year, as it was in the 1960s.

Case studies: the poor people

The analysis of any one particular family is becoming increasingly complicated, because of the population increase, the division of holdings, the links between several brothers—each having inherited part of the land,[7] but some having gone away, with or without their family. The story of landless people is not simple either, in view of a combination of several jobs, of people being outside or not.

Ajit is a Bhangi, rather young, not too energetic. Every day he sweeps the courtyards and homes of three Jat families. Each of them gives him one thick *chapati* every day and around 30 kg. of grain for six months. Ajit is not the poorest of his community. He owns two sets of clothes and has acquired a small buffalo from a cultivator. He is responsible for feeding it and when it starts giving

[6] A friend and colleague, Jean-Luc Maurer, who visited Mirpur after making very minute village studies in Java was surprised by such high expenditure compared to what he had observed in Indonesia.

[7] Joint families were already breaking down in 1963, after the death of the father, a trend even more common now.

milk, he is allowed to keep half of it. He has to maintain his wife and his five small children, through additional odd jobs. During the wheat harvest he earns about 10 kg. per day.

Saggan, another Bhangi was doing quite well in 1963. He was raising pigs and had a few chickens. His only son was working in Delhi. In 1975, Saggan's life was still not too bad. He had no more pigs, the chickens had died, but he was relying on three goats. Later, his son lost his job in Delhi and returned to Mirpur to work as a sweeper for several families, but he is now in poor health. When we met Saggan in 1986 he was an old man, weak and very thin. Following the death of Tez Pal he lost his job. He has also lost his self-confident nature, the one thing that struck me about him in 1963.

Dalchand, also a Bhangi, has been doing fairly well. He had spent six years in the army in Bombay. By 1978, he was driving Tez Pal Singh's tractor. He had a buffalo which gave 3.5 litres of milk per day, 2 litres of which he sold. Two small buffaloes were soon to produce milk. Borrowing money from the Punjab National Bank, he was able to buy pigs for Rs. 700 and will sell them for Rs. 1,000 a year later. He also suffered from Tez Pal Singh's death since he lost his job as a tractor driver. Occasionally he repairs one. His pig business is going on and he manages to get a net profit of Rs. 900 after one year. He was worried about his buffalo which no longer gives milk. Was it sick or pregnant? He also gets some work on a *kolu* and does some other odd jobs occasionally. Dalchand has a very clear mind and understands the weaknesses of his community, such as poor education, very low status even among Scheduled Castes,[8] sometimes lack of an enterprising spirit. He also complains about Jats and others who are now cleaning their homes themselves. Dalchand is no less clear headed about the government schemes for the poor such as the loan-cum-subsidy. Although he benefits from these schemes, he wonders whether government expenditure could not be used in a more employment-oriented way, for instance, for better infrastructure.

Haryan Singh represents a good example of those Jativs who are today better off or certainly less poor than they were twenty years ago. He managed to go to school and got a scholarship to pass his matriculation. Aged 45, he is alert and works mostly as a

[8] There is also a hierarchy within the Scheduled Castes in terms of social status.

mason, earning Rs.20 per day. His buffalo gives 5 litres of milk per day, 4 litres of which are sold. He has three boys, the elder one going to school in Mirpur, and one girl. He owns two sets of clothes, has bought a bicycle and even a watch. 'It is true my life is better now, he concludes, but the *babus* (bureaucrats) make more money.'

Dala, another Jativ, is 30 years old. It was only after his wife had her fifth child that she was operated. Dala does casual agricultural work for which he does not always get the usual rate of Rs.10 per day or, more rarely, Rs.12. It may be Rs.7 or 8, but with some food. He has no semi-durable goods and no extra clothes in addition to the ones he wears. His small buffalo should soon give milk. Dala has benefited from a loan-cum-subsidy of Rs.4,000 to buy a buffalo and *bogey*. The deal was, however, not so smooth. The veterinary pocketed a *baksheesh* of Rs.60 and the clerk of the bank Rs.300. He goes to the Ganges to bring back earth (Rs.20) or to Jahangirabad with a load of *gur* or other commodities, getting Rs.30 (10 quintal load). Every month he has about ten days of work in transport.

Nandram, a Jativ, does reasonably well through the same activities as in 1978-79. As a mason he earns Rs. 20 (Rs. 10 in 1978-79). He also earns some money for brick making and in agriculture he receives Rs. 10 per day. In 1978-79, his buffalo was giving 5-6 litres per day. He had three small buffaloes. Now Nandram has lost one young buffalo and the two other young ones may have been sold. Nandram does not sell milk, but sells *ghee* (clarified butter) made at home. The four boys are now teenagers, the girl should be married within a few years.

Bashir is a Muslim who used to be a permanent labourer of Tez Pal Singh. In addition he also had an oil crusher (mustard) and one bullock. In 1963 he complained of stagnant living conditions. He has taken advantage of the rising traffic of goods. In 1978, he bought a *tanga* and a house with a loan-cum-subsidy (total Rs.2,500). Every morning he loads the *gur* (raw sugar) manufactured by Ram Singh during the night. He carries 360 kg.of *gur* for Rs.18. He then takes Rs.2 as Ram Singh's fare who delivers the *gur* at Jahangirabad 15 km away. He takes additional passengers (also for Rs.2) and hopes to find customers for the return journey. To feed his horse he spends Rs.15 per day, which does not leave much profit. He complains of the rising competition in transport.

Occasionally he is able to find some additional work. All in all he manages reasonably well. His four elder sons are working and living in Delhi with their families. From time to time they send money home to their father who still has three younger boys and one small daughter to support. Another daughter is married outside, and Bashir has to repay a debt of Rs.2,500 (at 36 per cent interest) borrowed for the marriage.

Kishora, a Jativ, is on the whole doing better. In 1980, he opened a flour mill and a shop where he sells cigarettes, onion, pulses, grain, soap, batteries and biscuits. For that he got a loan-cum-subsidy of Rs.4,000. One of his sons is working as an agricultural labourer in another village. Another son is at school, and the two youngsters stay at home. Kishora is one of the relatively few Jativs who now lives in a brick house, to which he is adding a second floor. He is quite happy with his improved living conditions.

I met Maqbool again, a Muslim[9] and a carpenter by profession. In 1963 he was one of the poorest of the poor, i.e., a little above survival. He used to make and repair agricultural implements for ten to fifteen families, each giving him 37 kg. of grain per year in return for his services, a typical example of the old *jajmani* system. He also worked in the field or rented out his bullock. His buffalo had just died. Until then, he was making *ghee* (clarified butter) for sale. Maqbool is now an old man. He has less work repairing bullock carts, because of the growing use of tractors. He still works as a labourer for Rs.10, whenever he finds an opportunity. Fortunately, one of his sons has found a job in a small furniture factory in Delhi. He saves Rs.300 per month which he sends home to his young wife. The second son is also working in the factory as an apprentice, an example of the adaptation of traditional work to a modernising economy.

Jaipal Singh who died some years ago was a poor Jat without any land. In the 1960s he worked as a tailor in a tiny shop where he also sold some cotton, *beedis* and matches. In those days he was paid more in grain than in cash. Unlike many Jats he did not have an enterprising spirit. His son, Manohar Singh, who has replaced him, is showing more ability. He has widened his business and bought a small cotton ginning machine. He buys cloth and makes

[9] There are about 150 Muslims in Mirpur and I never heard of any tension or trouble between them and the Hindus.

razais, a kind of double blanket stuffed with cotton. He also cuts mill stones for flour mills, using a 5 HP electric engine or a diesel one when electricity fails. In one month he earns a net profit of Rs.300. His wife looks after the buffalo which supplies 5-6 l. of milk per day which is consumed at home. One of his brothers works in Mirpur, while another is a policeman in Delhi where he lives with his family.

The landowners

Mohammed Sharif, a Muslim, and his brother Nur Baksh are now old men surrounded by a large joint family. Fourteen people, including wives and children, live here. Obviously, 1.2 ha. of land are not enough. Two-thirds of the area is devoted to wheat following fodder crops and maize grown in the monsoon. Sugarcane occupies the last third. Three buffaloes are fed close to the house. Out of a daily output of 12-15 litres, 8 litres are sold at Rs. 2.20 per litre. This large household is not badly off, because two young men, in spite of being educated only till the primary level, are working in a New Delhi factory making steel structures. Each of them earns Rs. 1,400 per month. One of them sends Rs.400 each month, whereas the other sends less because his wife and children are with him.

Sukharam Khatik, had already lost his wife when I first met him. He was the head of a big household. His three sons were all married and together they had seven small children alive, twelve had died in their early years. All these people could live frugally on 4 hectares. They used to sell 370 kg. of wheat and consume 1,160 kg. per year, plus maize. They had two bullocks, one camel which was used on the *kolu* and two milch buffaloes. Part of the milk went to the children, the rest was sold. Sukharam was not too clear when I put my usual question to him: 'Is your life better now or not?'

What is the situation today? After the death of Sukharam each of his sons inherited 1.3 ha. Girraj, one of the sons, is 50 years old. His two daughters are married and his only son is still young. With this small family[10] Girraj is pretty well off. He sells the cane grown on 0.5 ha. On 0.75 ha. he harvests 2,000 kg. of wheat, keeping

[10] He lost four children.

1,200 for his home requirements and selling 800. *Jowar* and *bajra* feed his pair of bullocks and his two buffaloes, each giving 5 litres per day, half of which is sold. half consumed. Yet, one should also take account of the heavy expenditure incurred in marrying his daughters. For the second one he spent Rs.20,000. Girraj has taken on on a sharecropping basis, 1 ha. belonging to an old Thakur. Both the cost of input and the output is shared. Girraj is also the co-owner of a tubewell. He has pulled down his mud-walled hut to construct a brick house.

In the compound live the two other brothers with a similar life except that Virandar's wife has started making dresses for sale (earning Rs. 150 per month) on her sewing machine.

In the same compound live another group of brothers who also belong to the Khatik caste. The five of them have each inherited only 0.12 ha. Haripal is in his fifties. He grows 400kg. of wheat followed by 240 kg. of maize. His two buffaloes yield 7 litres of milk out of which 4 litres are sold at Rs.2.0 per litre in Mirpur. Fortunately, two of his sons have found a job outside and send money home to help their younger brothers.

Let us shift to the quarters of the Lodha Rajputs. In 1963, old Hetram was a rather big farmer owning 7 hectares of land, but he had a very large family of four sons, all of whom were married in the village and had, at that time, sixteen children in all. Only one of the men was employed in another village as a Village Level Worker. By that time they were selling all their *gur* and a few bags of wheat. They were getting a good supply of milk from their buffaloes which was consumed at home. They had four bullocks and two camels.[11]

After Hetram's death, each brother inherited 1.75 ha. In 1979 Nawat Singh, one of the brothers had three girls and two boys at school. The eldest son had gone to Kanpur with his family to work as a mechanic in a mill. Nawat and his brother jointly owned a tubewell. Clever and hard working, they took the maximum from the green revolution. In 1978-79, Nawat was already harvesting 4,000 kg./ha of wheat and could sell 1,200 kg. His sugarcane (one-third of the area) was entirely for sale. They had two bullocks and two buffaloes supplying milk for the family.

In 1985-86, we met Nawat again. He had aged a little and was

[11] The few camels in the village disappeared after 1970.

walking with a stick. He complained of some pain in his back, but he remains as lively and full of common sense as ever. Another son has got a job as a mechanic in a mill in Panipat (Haryana) where he lives with his family. The other sons are now grown up and work on the land. One is married, the youngest ones are not. Nawat has reduced his wheat fields slightly, growing some more fodder. He now has three buffaloes. He also cultivates potato on 0.6 ha. between October and December. While he gets a good profit, this affects the yield of wheat,[12] sown just after. He sells his cane and potato in Jahangirabad .The whole family continues to do quite well. They now have a brick house, but they have few semi-durable goods. Yet, the future will not be easy: wheat yields, already high in 1978-79 have not increased since then. New and richer crops are needed and more jobs outside the village.

Mishrilal, a Brahmin, comes, like so many others, from a family of medium farmers. Like his two brothers, he has inherited only 1 hectare, including 0.2 hectares of mango orchard. His is a good example of a family increasingly involved in business. Two sons have opened shops in Unchagaon. He himself and his third son look after the fields. On nearly half a hectare they harvest 1,700 kg. of wheat. Then they push *berseem* on 0.3 hectares. With *jowar* and *bajra,* they get enough fodder to now feed six buffaloes. Every day, 15 litres of milk are sold in Unchagaon. The orchard is rented to a businessman for Rs 1,500 per year. Mishrilal has added a mechanical *kolu* to his tubewell and he buys cane at Rs. 240 to Rs. 280 per t. and sells *gur* at Rs. 320 per quintal. Although he lives fairly well, Mishrilal complains about the rather small profit that he makes on his *kolu.* Besides, his son Omprakash already has five girls and one boy, which means a very heavy marriage expenditure in the future.

The case of Khanchand Sharma, a Brahmin, involves a new type of family arrangement. Once one of the local leaders, Khanchand is now old and sits in front of his *chaupal* or takes a short walk, leaning heavily on his stick. The whole family has succeeded very well out of 3.3 ha., following the usual pattern: a tubewell, new seeds, chemical fertilisers. The mud house has been replaced by a two-storied brick house. The *chaupal,* where we stay, has been enlarged. All the four sons have work outside agriculture.

[12] See Chapter 2.

Ram Kumar (my informant in Mirpur) is the general manager of the Inter-college of Mirpur and also acts as personal assistant to an important member of Parliament from Unchagaon. Harcharan teaches in the high school at Unchagaon. The other two are in Bulandshahr, one as a teacher, one as a clerk with the Food Corporation of India.

The land has been divided equally between the four sons. Ram Kumar looks after his own fields with a permanent labourer. Harcharan takes the fields of his two other brothers on rent. He employs a labourer and works on the land himself when he returns from Unchagaon. The whole family lives comfortably, enjoying not only better food and accommodation than in the 1960s but also the usual semi-durable goods like bicycles, watches and transistors.

The Jats, like others keep pushing their main crops and diversifying them. Sauraj Singh is one of the four sons of a relatively big landowner, Ramachandra Singh who died in 1980, leaving to each son 1.25 ha. His two sons are here, including the eldest one who is married with several children. There are also two young daughters. On 0.75 ha. he grows wheat with a yield equivalent to 3,200-3,800 kg./ha. This year (1985-86) the sugarcane yield fell severely. Sauraj Singh usually rents a tractor (he has only a small bullock) for Rs.640 per ha. for tillage operations. He still manages to sell some wheat along with all the sugarcane. He keeps all the milk from his two buffaloes for his household. Near the house, Sauraj Singh has started growing potato and tomato followed by other vegetables.

At the moment the family is doing reasonably well, but one can imagine their fate if, as in the time of Ramachandra, they were growing about one-third of the wheat presently harvested. Yet, the future will not be easy. The marriages of the two girls will be costly and at least one of the young sons, if not both, will have to find a job outside.

Tikkam Singh's household (another Jat family) has also been broken into four after his death, leaving to each son 1.6 ha. One of them, Girraj Singh, has been particularly clever. He has helped his elder son Sripal to buy a second hand truck (Rs.20,000)[13] to start a business in Jahangirabad where he stays with his family.

[13] He first sold some land and then bought other plots later on.

The two others, around 20 years of age have been educated. They are typical of the new generation of people not keen to work on the land. One of them is married and received, among other things from his in-laws, a television set valued at Rs. 3,500.[14]

Girraj Singh also earns money through the sale of water from his tubewell to some neighbours as well as by crushing cane in his mechanised *kolu* for other farmers. Around the house several papaya trees were planted some years ago. If the two educated sons do not find a job, the family may find themselves in trouble, especially with the birth of grand children.

It is always a pleasure to meet Ram Singh who has come a long way since the time when he discussed the installation of his tubewell with me, following the consolidation of holdings. In 1968, he stood proudly in front of it, surrounded by beautiful wheat fields. Ram Singh had to work hard on 4 ha. to feed and educate his four sons and two girls. The three older sons are away, two work in Delhi, one as an irrigation engineer and one son works in Jaipur as a sanitary inspector. The fourth one is in the village. He already has two boys and three girls, but is reluctant to undergo an operation.

For the last eight years, the wheat yield has remained around 3,200 kg. ha. Ram Singh is using new mustard seeds which give a good crop of 960 kg. for 0.5 ha., sold for Rs. 3,600. Ram Singh also devotes 1.6 ha.to *berseem* from October to April. Part of it is consumed by two bullocks, a buffalo that pulls the *bogey* and two milch buffaloes. Ram Singh has installed a mechanical cane-crusher on his tubewell. All in all, Ram Singh is diversifying his farming, reducing wheat in favour of other crops. He can face the future with confidence, especially if his fourth son manages the land in the same shrewd way as he himself does.

Towards the future

The present situation may remind us to some extent of the conditions prevailing when I first left Mirpur in August 1964. At that time the old patterns of production and employment possibilities

[14] There are five television sets in Mirpur. But the poor supply of electricity does not allow people to fully enjoy them!

were approaching their ceiling at the very moment when the population was increasing faster.

Even though there has been no decisive fall in the birth rate with the result that the population has increased by 50 per cent in twenty years, the economy has overtaken the population in all fields. Basic agricultural production has considerably increased, jobs opportunities outside the village have more than trebled; jobs outside agriculture in Mirpur itself have also expanded.

At this stage, I wonder whether Mirpur will not be forced to enter a new phase. Since 1982[15] the economy seems to have been losing some of its momentum. Wheat yields have levelled off. Sugarcane yields are not increasing for lack of new varieties. The diversification of the economy: *berseem* fields, the relative come-back of mustard and potato, do not yet play a substantial role. As for milk, how much can it expand under the existing conditions (in terms of the availability of fodder and cost of cattle).

Small business and transportation could also face difficulties. Typical are complaints of *tanga* owners about rising competition. As for the construction boom, it is slowing down, because most of the people who can afford to have by now shifted from mud to brick houses.

We have seen how important the migration of workers outside Mirpur has become, a factor which helps so many families without land or with holdings below 1 ha. to nevertheless lead a decent life. It is difficult to foresee whether or not such safety valves will remain as flexible as they are now. The rising number of educated young men, more or less idle in Mirpur, is a matter of concern.

What could be the new agents of change, bringing Mirpur some more steps ahead (apart from migrations to larger villages and cities)? It is to be hoped that family planning will at last have an impact on a large scale, but even if it happens it will not help much in the short run in the field of employment, because the future workers are already born and they are in great number.

The first source of further development would be a nearly perfect supply of electricity. This would lead to an equally quasi-perfect system of irrigation, which would help boost yields of wheat, or *kharif* crops and sugarcane further. Such a system would also enable farmers to replace part of their present *kharif* crop by

[15] I spent a few days in October 1982 and April 1984 before the 1985-86 survey.

a crop giving a higher return like rice, as has been done in Punjab. For the whole of Punjab the average yield of rice followed by wheat is 3,000 kg./ha. per crop.

A smooth electrical supply would also reduce the time needed for threshing wheat with the present machines. It could help farmers to push more pulses (*mung* beans) between the wheat harvest and the advent of the monsoon crops. Pulses fetch a high price, because their supply all over India has fallen short of the demand.[16]

Water and electricity are not the sole factors which could boost production. Many farmers should use a better balance of fertilisers between urea and DAP in increasing the share of phosphatic fertilisers and also some more of both or a more efficient use of the same doses. A more sophisticated combination of chemical fertilisers is also needed. When adding some zinc and copper, one may gain six more quintals of wheat per hectare. At the moment it seems that the renewal of wheat seeds, every four to five years, does not work too badly, but there is room for improvement. As regards pesticides, not much is needed for wheat, but it is necessary for sugarcane and for monsoon crops.

All these tasks are well within the ability of the farmers. They would then reach 4,000 to 5,000 kg./ha. of wheat, followed by 3,500-4,000 kg./ha. of rice. But they are not their own masters. They depend on outside factors for the supply of inputs. For several years the greatest constraint which blocked most other factors of growth was electricity. As will be seen later in this book, one cannot be too optimistic until 1990 or even later.

It is understandable that a partial improvement in the supply of electricity would affect yields, but, of course in a limited manner only. The point to bear in mind is that the higher yields quoted above will be very demanding on several sectors of the economy.

As for the diversification of the economy, part of it could be achieved under the present power supply, but again, it is not so simple. A further expansion of potato production would increase the area devoted to late sowing of wheat and thus reduce the yields of the latter, as seen above.

[16]The production of pulses has been fluctuating without a decisive upward trend for the last twenty years. Pulses also enrich the soil by fixing nitrogen from the air.

Another factor of growth is transport and communication. The network of all-weather roads is about to be completed which will boost traffic and trade of all kinds. How far will it affect new small industries? It is difficult to answer this because such industries will be affected by power shortages.

To conclude, the immediate future of Mirpur is not entirely secure, like that of similar villages in the area. The next steps may not be easier than the ones reviewed here and it is here that the problems of the youth come up again. There is a risk that the new generation may not be of the same sturdy and hard working calibre so well illustrated by their fathers.

Chapter Four

The District of Bulandshahr

Let us review the achievements in Unchagaon block and Bulandshahr district over the last twenty-three years, as well as the basic features of government policy.

The Unchagaon block

The basic data[1] are as follows:

Geographical area, 146 sq. km.
Population (1981 Census), 104, 604
Density per sq. km., 706.

Mirpur fits fairly well into that pattern with a density of 669 per sq. km., a little below the average for the block. What is surprising, however, is that the population has only increased from 77,000 to 104,604 between the two Censuses of 1961 and 1981. It seems a trifle on the low side, even when acknowledging that a rather large number of people migrated outside the block, confirming the trend observed at Mirpur but probably with a higher proportion of migrants.

1985:	Net cultivated area		12,837 hectares
	Gross cultivated area		26,999 hectares
	Net irrigated area		12,587 hectares
	Of which:	canals	4,620 hectares
		tubewells	7,643 hectares (7,328 by private tubewells, 315 state tubewells)
		wells	262 hectares
		other sources	62 hectares
	Gross irrigated area		22,671 hectares

Persian wheels have practically disappeared in favour of tubewells,

[1] All the data are from the block records.

mostly privately owned, but canals which hardly touch Mirpur are more important in other villages. Practically the whole area is now irrigated. As for public tubewells, originally they played an important role, but they are no longer encouraged and often are not properly maintained. Some, as in Mirpur are not working at all any more.

	Main crops (ha.)		*Yield : kg./ha.**
1984-85:	Wheat	9,810	2,623
	Maize	5,955	1,459
	Sugarcane	5,503	4,704 (in terms of *gur*)

*rather rough estimates

Barley, peas, *arhar* (pulse) each cover only a few hundreds of hectares.

The intensity of cropping has grown considerably. The index of multi-cropping which was equivalent to one-third of the cultivated area in 1961 is now above 100 per cent which means that several fields seen in Mirpur bear three crops per year — *jowar* or *bajra*, potatoes and wheat — while cane is a one year crop. Average yields tend to be lower than in Mirpur, a rather normal phenomenon since in some villages irrigation is below Mirpur. As regards cane, one has to bear in mind the unusually low yield due to pyrilla attacks. Normally it should be above 5,000 kg./ha. New seeds and chemical fertilisers are easily available and used in a manner similar to that in Mirpur. There are now 325 tractors covering about half the cultivated area. The progress of milch buffaloes is a common feature all around but there are also now 100 cross-bred cows.

The number of cultivators is 16,390 out of which 10,000 cultivate below 1.3 ha. There are about 4,250 landless agricultural labourers. Again, a pattern into which Mirpur fits. Domestic small industries and handicrafts (potters, brick making, carpenters and others) provide employment to 1,779 persons. Construction, trade and other activities involve 1,460 people (1981 Census). Small industries in the village of Unchagaon now have eighteen units employing seventy-nine workers. These figures do not include repair workshops and trade.

In 1963-64, and not infrequently even later, I was fairly unimpressed by the block activities: agricultural extension, credit and overall planning and development. This time the performance was better. The block development officer (BDO) is very smart, quick and efficient. He knows and understands his facts and figures. He has a good rapport with peasants who speak highly of him and of his integrity. The team of extension workers and others around him are not of the same calibre, but, working under such a person, they achieve more than when directed by a rather mediocre, *mamuli* BDO.

At the district level[2]

Coming to a larger area, one observes more diversity than within a single block. In the past twenty years the few towns have grown considerably. Bulandshahr keeps expanding. The old bazar is more and more congested. New ground on the left bank of the river is now full of workshops, shops and bus stands. Near the highway new colonies are being built. Jahangirabad, as mentioned in Chapter 2, is also growing fast. According to the 1981 Census data for the whole district:

Total labour force	610,352
Rural	494,761
Urban	115,591

The process of urbanisation increased in tempo after the advent of the green revolution.

Percentage of urban population	
1961	12.6
1971	13.3
1981	19.3

The data should be taken as a rough indication because the district lost some villages and small towns between 1971 and 1981.

The rural areas surrounding the cities were already affected in the early 1960s when wages of agricultural labour were higher than in remote villages like Mirpur. Such a trend accompanies the expansion of vegetable farming.

On the other hand, some rural areas (almost 5,340 ha.) suffer from alkaline soils and are falling out of use. Weak irrigation in

[2] All data have been supplied by the district administration.

some areas may also bring down the average yield for the district as compared to Mirpur.

Geographical area (sq.km.)	4,352
Total population (in millions, 1981)	2.36 million
Density per sq. km.	542
Net cultivated area (ha.)	341,066
Gross cultivated area (ha.)	589,489
Net irrigated area (ha.)	327,386
of which : canals	72,832
tubewells	241,665
traditional wells	7,744
others	6,145

Fifty-eight per cent of the net cultivated area is multi-cropped and irrigation now covers practically the whole district as against 70 per cent in 1961. Besides, the quality of irrigation has improved considerably with the replacement of Persian wheels by tubewells.

Main crops (ha.) (1984-85)		*Yield:kg./ha.(1984-85)*
Wheat	222,400	2,560
Maize	116,000	2,360*
Jowar, bajra	35,000	–
Sugarcane (*gur*)	49,000	4,000–5,000
Rice (clean rice)	5,500	926**
	12,500	–
Mustard	7,300	750

*Until 1984, around 1,500 kg./ha. This sharp increase is due to the introduction of new (non-hybrid) seeds. It will be interesting to see if this upward trend continues.

**A very low yeild, may be due to the recent introduction of rice or other factors. (The average yield in Punjab amounts to 3,000 kg./ha.)

The consumption of chemical fertilisers has also risen tremendously.

1960-61	6,890 t. (N P K, in term of nutrients)
1968-69	12,300 t.
1978-79	32,600 t.
1980-81	55,900 t.
1984-85	61,610 t.

However, the ratio between nitrogen and phosphatic fertilisers remains inadequate with not enough of the latter, which may in the long run affect the soil.[3]

Out of 244,000 landholdings, 137,000 are of less than 1 hectare, 52,300 are 1 to 2 hectares, 44,400 from 2 to 5 hectares and 10,200 more than 5 ha. (round figures). In 1981, there were 315,400 cultivators versus 111,300 agricultural labourers (census data).

While industries, except sugar mills, were rare in the 1960s, there are now ninety-five registered factories employing 6,800 workers making synthetic fibres, foam, paints and allied products, paper and carpets. The number of sugar mills has risen. There are also 1,265 small scale units (each with less than 10 workers) employing 5,970 workers: mechanical and electrical workshops to attend to tubewell equipment, tractors and trucks, bicycles, transistors and the manufacture of pumps.

Trade is expanding no less with many shops supplying semi-durable goods. Bulandshahr now has a few cinema halls and modern restaurants as well.

Rural development policy

In the 1960s and even later, district planning and the overall performance of the administration was not very striking. In 1986 it was much better. The senior cadres in charge of the various departments—agriculture, irrigation—were better informed and more experienced. The associate district magistrate was in charge of the overall operations. The collector, who previously played a rather distant role in rural development is today much more involved. The young IAS (Indian Administrative Service) officer in charge is both energetic, imaginative and seems very committed to his task.

We proceeded in our survey with a wide exchange of views on the district. Already fairly advanced in the 1960s, though less than Meerut, Bulandshahr remains a progressive district. Yet, the future phases of development will not be easy. Several blocks should benefit from the new Madhya Ganga Canal under construction from Syana to Jahangirabad. It will bring water as a complement to tubewells, which should induce farmers to boost

[3] 1984-85 *Kharif:* 16,740 t. of N, 1,100 t. of P_2O_5, 450 t. of K_2O.
Rabi: 30,400 t. of N, 9,760 t. of P_2O_5, 3,180 t. of K_2O.

paddy cultivation instead of maize. There will also be more fodder available like *berseem* and lucerne.

As far as wheat is concerned, the assistant director of agriculture rightly emphasised the need for a better balance of NPK, a realisation that was rather slow in coming considering that the fact has been known for so long. Early sowing (i.e., before the end of December) is recommended because the shortage of power for irrigation is often acute in March. It can get worse in the following months and affect sugarcane production. Potato cultivation has become widespread but could increase further. Maize remains vulnerable due either to excess rain or drought, although the new seeds look encouraging.

The authorities are propagating afforestation as seen in Mirpur. In 1985-86 they distributed 8 million saplings of eucalyptus for Rs. 0.10 per piece (actual price Rs. 0.25; 0.15 as a subsidy).

We discussed at length the various schemes designed to help the poor, particularly the Integrated Rural Development Programme (IRDP). The authorities are well aware of several pitfalls (wrong identification of the poor, *baksheesh* taken by various officials) and of other difficulties, such as a loan-cum-subsidy for buying a buffalo which later on, the buyer has to sell because he canno maintain and feed it properly. The people themselves have mixed feelings towards these programmes complaining of all sorts o malpractices.

On the other hand, as seen in Mirpur, there are some positive cases where the poor have really benefited. The impossible task i to assess, in a relatively precise way, the positive and negativ aspects of the target group approach, or, what one can call th direct attack on poverty. The other approach, the indirec approach, does not involve special measures for the poor, but the do benefit from the global process of rural development of whic Mirpur is a good example.

The IRDP alone has cost (during the Sixth Five-Year Plar 1980-85), Rs. 4 million in subsidy and Rs. 8 million as loans fc each of the seventeen blocks. Should one expand such progran mes or should one give more emphasis to the indirect approacl or, what other kind of blend could one imagine ? These sam debates will come up again in all the districts visited, later on in th state capitals and finally in New Delhi when it will be time summarise our observations.

We concluded our discussions with the subject of family planning. The senior district officers do not attempt to hide their worries. One cannot resort to coercion, especially after the excesses committed during the Emergency (1975-77) which was one of the main causes of the debacle of the Congress Party in the elections of 1977. On the other hand, a more rapid decrease in the rate of population growth is needed very urgently.

A new scheme has just been introduced in UP. When either spouse is operated (sterilisation) after the birth of two children, the couple is entitled to a green card which carries various advantages for finding a job, buying certain items like kerosene and obtaining a loan from a bank. Earlier, sterilisation entitled a man to Rs. 40 and a woman to Rs.100 and sari. Will the new incentives be strong enough ? How will they operate in the villages where the problem is more acute than in the towns?

Politics, administration and the peasants

Regardless of whether they are literate or illiterate, poor or rich, the villagers have become increasingly sensitive to politics. At the same time, they are more and more critical of the way many politicians behave, a phenomenon by no means confined to western UP, but noticeable practically everywhere. Patronage, undue privileges and corruption are hotly debated topics and farmers can be very blunt even when talking to senior officials. I saw how my friend Ram Singh complained to a minister of the UP government of the adulterated chemical fertilisers.

On the one hand, I came across a number of efficient and dedicated civil servants who were respected by farmers both at Unchagaon and Bulandshahr. On the other hand, peseants complain more than before about the deterioration of public and social life. Corruption, especially at the lower echelons of clerks and policemen had always been there, in British days as now, but the disease is spreading in many directions. If you apply for a job in the police, a tip is needed. Even for recruitment in the army malpractices are not unknown. A teacher told me that unless he pays some money to the bank employee, gets his salary cheque only after considerable delays. And then, when I ask, 'why do you not complain,' the answer is *'koi nahin sunta'* (nobody listens to us). I have heard these very words since the beginning of my village studies in the 1960s, but they come up more often now. Even in 1978-79, these complaints were less common. At that time, as in the 1960s, malpractices were usually associated with the purchase of rationed goods or cooperative loans.

The break down of law and order is not confined to Unchagaon block. Several burglaries took place even in Narora, on the site of a new nuclear power station. As for problems concerning youth, district authorities do express their worries.

In comparison to Bihar, the situation remains quite peaceful here, but again, in the past few decades one heard of criminal offences much less than one does now.

Chapter Five

The Rice Bowl of Tamil Nadu

We enter now a vast coastal region where the river deltas with their good alluvial soils have enjoyed a very ancient tradition of canal irrigation systems. The earliest canals in the Cauvery around the old city of Thanjavur (Tanjore) can be traced back to the 2nd century AD. The tradition of large tanks in the interior is no less ancient, maybe even more so.

Referring to the Pallava and Chola dynasties 7th-12th century AD, Burton Stein writes of 'a large number of localities of peasant society and culture [with] four major clusters : in the Kisna [Krishna] delta, ...in the basins of the Palar and Ceyyar rivers, the Ponnaiyar basin and the Cauvery Basin'.[1] While Thanjavur has remained until today the main rice bowl of south India, the Krishna delta (Andhra), where Guntur district is located, went through severe ups and downs. During the first part of the 19th century, 'the canals of the Cholas had been long choked or forgotten in the delta.' Irrigation was confined to a few tanks and wells. Paddy was confined to small plots of land. Millet and pulses were the dominant crops, bearing low yields. When nature turned hostile (through droughts or cyclones) conditions became worse. In the 1833 famine, for example, more than half the population was wiped out.[2]

In both areas as well as in the Godavari delta, the British created large new canal systems while they were enlarging the irrigated area in the Cauvery basin. Resuming the tradition of the Cholas, Sir Arthur Cotton built barrages at the head of the Andhra deltas in order to feed a wide network of canals. New works were also added on the Cauvery, especially (in 1935) the Cauvery Mettur Project (CMP), a large reservoir with a dam, which enabled the irrigation of the so-called new delta where I made my surveys.

[1] B. Stein, *Peasant, State and Society in Medieval South India,* (New Delhi, Oxford University Press, 1980), p. 67.

[2] See R.E. Frykenberg, *Guntur District 1788-1848,* (Oxford, Clarendon Press, 1965), pp. 2-8.

The progress of irrigation boosted paddy cultivation and, in the case of Guntur, tobacco. Roads, railways, market towns, trade, some industries, rice mills, tobacco factories and workshops expanded as well.

This early global process of rural development went several steps further ahead after Independence, so that at the time of our first surveys, both districts were much advanced compared to many others. However, such a process seemed more advanced in coastal Andhra. Touring east and west Godavari districts in 1966, Daniel Thorner was already struck by their affluence: 'even the families of agricultural labourers earned enough to eat three rice meals per day supplemented by vegetables, spices, chillies, and occasionally dried fish and tea... Their houses were [a] world apart from the wretched shanties I had entered in North Bihar.'[3]

While these deltas offer certain similarities, there are also substantial differences with reference to the actual situation and the future.

The village of Kila Ulur, Thanjavur district[4]

Until 1935, there were practically no canals in the new delta. At the time of the last settlement (1924) only 58 ha. in Kila Ulur were irrigated by tanks and wells, as against 309 ha., rainfed. At that time paddy covered 103 ha., groundnut 160 ha., and the other fields bore various kinds of millet.

After 1935, the command area of the canals covered 333 ha. of paddy; non-irrigated land fell to 36 ha. to which one should add small plots irrigated by temporarily dug wells in the dry season.[5] The region then entered an entirely new phase in its history, becoming similar to the old delta. From then on, the cultivated area remained unchanged as there was hardly any land to reclaim.

The population increased very slowly from 1,310 to 1,444 between the Censuses of 1901 and 1951. Since then, the rate of increase has gone up, though moderately, so that in 1981 the population was 1,800. Such a trend is rather typical of many parts

[3] D. Thorner, Coastal Andhra Towards an Affluent Society, *Economic and Political Weekly*, Annual number, February 1967.

[4] In Orathanad *taluk*, located 15 km. from Thanjavur.

[5] One must add around 80 ha. for the different hamlets and casuarina woods.

of Tamil Nadu, the explanation for which is not totally clear. For instance, between 1951 and 1961, Tamil Nadu's population grew by only 1.5 per cent per year compared to the all-India average of 2.15 per cent. Until the war of 1939-45, there had been substantial emigration to Southeast Asia and Ceylon, and also immigration to other parts of India. While the latter trend has continued, the flow of Tamils outside India stopped after 1939. A number of Tamils came back after 1945.

As regards natural fertility, it has been relatively low for several decades. In 1964 I came across a number of couples in Kila Ulur who had only a few children.[6]

During the last decade (1971-1981), the difference between Tamil Nadu and the all-India population increase was no less striking : 17.23 per cent as against 24.80 per cent. During this period, family planning must have had an impact because it has been much more successful here than in a number of states like Uttar Pradesh or Bihar. In Kila Ulur, family planning clearly fared better than in Mirpur.

What have been the trends regarding land tenure? The old raiyatwari system remained practically unchanged. In Kila Ulur, the four main landholders have between 6 and 8 ha. of land, so that if the ceiling (6 ha.) had been implemented, one would have recovered hardly any surplus land to distribute to the poor.

Comparing data for 1964 and my last visit in 1985, one finds a trend rather similar to the one observed in Mirpur : the total number of holdings has nearly doubled. Now 180 holdings are below 0.8 ha., 50 between 0.8 and 2 ha.

The Kallar represent the dominant caste, owning the majority of the land. They are good and clever farmers. As soon as the canals of the CMP reached their village they adopted the classical paddy system of Thanjavur : *kuruvai* (105 days), end of June to end of September, followed, on the same soil, by *thaladi*, end of October to February. The largest part of the fields was devoted to *samba*, a crop of longer duration (145 to 180 days), from August to January.

In the early 1960s, the Kallars and the Ambalakars (the second main agricultural caste) had reached typically high yields within a

[6] All these comments are tentative. Such issues deserve a detailed study by demographers.

traditional framework: careful land tillage, early growth of paddy in a well kept nursery, transplanting, use of some farm manure and weeding. Irrigation was far from perfect even then, but it was more or less adequate for the existing crop pattern with an average yield of 2,000 kg./ha. of paddy.

In 1961, Thanjavur had been selected as one of the districts where the Intensive Agricultural District Programme (IADP) was introduced. Sponsored by the Ford Foundation, the programme aimed at improving extension services, supply of inputs and credit, in order to boost production. Though not revolutionary in its impact, the IADP was beginning to be felt when I came to Kila Ulur in May 1964. The consumption of chemical fertilisers was rising. Some farmers had reached nearly 3,000 kg./ha. of paddy per crop and one of the largest landowners could, in a very good year, reach 4,000 kg./ha

As for the other crops, there was some groundnut (850 kg./ha., unshelled) some *ragi* (millet) with a yield of 1,000-1,400 kg./ha. and black or red gram (500-600 kg./ha.), cultivated after rice.

Even in those days farmers and government servants were concerned that irrigation could not cope with the requirements of higher yields. Delays in the monsoon, with the result that the Mettur reservoir would not be full in time, would prevent higher yields. Second, by January there is no more water available for irrigation, which prevents a second crop after *samba* which, at that time, covered three-quarters of the paddy fields. Then, during the monsoon, one can run into trouble if the rains are too heavy (floods), or so weak that the reservoir gets empty prematurely.

These problems are typical of all canal irrigation systems depending on the peninsular rivers. Unlike the Ganges and the other Himalayan rivers which depend on the melting snow *and* on the monsoon, the rivers in central and south India rely only on the rains. For that reason, perennial irrigation is rarely available as it is in the Ganges and Indus basins.

In the past twenty years, while there was no real improvement in irrigation, new varieties of paddy along with the rising consumption of chemical fertilisers, [7] have contributed to a sizable increase in production:

[7] Pesticides are needed much more during the monsoon than for irrigated crops during the dry season. This is because various kinds of pests are more predominant during the rains.

	1963-64	*1977-78*	*1984-85 (kg./ha.)*
Kuruvai	2,400	4,500	4,300
Thaladi	1,850	2,770	3,500
Samba	2,240	3,360	3,600

These data need qualification because they refer to rather good years. If drought affects the supply of canal irrigation, production may fall by 1,000 kg./ha. per crop. Then, in case plant protection is unable to fight pests like tungro virus, losses may amount to the same magnitude. In other words, the progress of paddy production has been both less significant and less steady from year to year, than for wheat in Bulandshahr district. Also, variations in yields have tended to increase with the advent of the new seeds and higher consumption of chemical fertilisers.

As far back as 1982, I noticed a new trend. Instead of growing groundnut on tiny plots irrigated by a temporary well in the dry season, some farmers had started replacing *thaladi* with canal irrigated groundnut, yielding around 1,200 kg./ha. (unshelled) and bringing a higher return than paddy. [8] There is also a trend to reduce the area devoted to *samba* in favour of two crops on the same soil.

In 1978, I had noted that four masonry wells equipped with a pumpset [9] (5HP engine) used to supplement canal irrigation and hence an additional crop in the dry season. In 1985, there was no new well-cum-pumpset. Tractors, unknown in 1964, are in common use by now, though some people still rely on bullocks. There are several people who own tractors now and there are also two power tillers. Both of them are also used for transport.

In 1978 a milk collecting scheme had been started. The metal road connecting the village to the main district road (a few kilometres away) had been completed, so that every day trucks could collect the milk. A number of poor people had benefited from a loan-cum-subsidy which enabled them to buy a cow. The scheme has now been discontinued: several people had no money to renew their livestock.

[8] India faces a serious shortage of oilseeds, hence the relatively high price of oilseeds in spite of large imports.

[9] A pumpset can be installed on an open well, not to be confused with a tubewell where water is pumped through a tube. A pumpset can also be set on a river bank or on a pond.

Other changes have fared better. One sees more pulses grown on the bunds of paddy fields. Since 1983, five farmers have started growing sugarcane irrigated by wells, especially in the dry season.

Vegetation around the different hamlets was always more luxuriant than in the north-west, a trend which continues today with more coconut, banana and bamboo trees in evidence. There are sales from time to time. One comes across more trees along the ditches or near the houses, particularly eucalyptus. Many families own a few chickens which they sell along with the eggs.

There were already some small shops in 1964. The number has increased, with a few coffee and tea stalls as well. The people have discontinued the practice of pounding paddy by hand to get rice (two-thirds of paddy). Their paddy is husked in a mill on the main road, a few kilometres away. Since the abolition of prohibition in Tamil Nadu, two shops selling toddy and arrack have now been opened.

Electricity appeared in 1965, but is used only for home consumption and for the few pumpsets. Farmers complain about the poor supply and the nearby transformer is not always in working order.

As elsewhere in Tamil Nadu, the local school supplies a free meal every day to the children. It appears to operate rather smoothly but there were complaints about the poor quality of food.[10]

A water supply scheme — a water tower with pipes connecting eight taps — has been completed at a cost of Rs.100,000. But, for many months it has remained idle for lack of some final connections. One wonders whether that type of system is really best suited. Could one not, instead, increase the number of private hand pumps?

The Paraiyars (the main Scheduled Caste group) have, for a long time, complemented their agricultural work by making mats out of palm leaves, an activity which continues today without a definite upward trend.

One comes across fewer new brick houses than in Mirpur. Many homes remain as they were with mud walls or partly plastered with bamboo and plaited palm and covered with a palm roof.

[10] In Tamil Nadu one often hears complaints of malpractices by the school administration and the teachers.

I cannot give a relatively precise estimate of the number of men working outside Kila Ulur. It seems that there are around 100 of them. Thanjavur is a growing city, a trade centre, with several medium and small scale industries. It was interesting to note that two Kallars and three Paraiyars have managed to go to Kuwait and Dubai.

While several trends in Kila Ulur remind us of Mirpur, the global process of rural development and economic diversification has been slower. Is it only because paddy has not increased as much as wheat has in the north? It is difficult to be sure because paddy plays a much more important role in the local economy than does wheat in Bulandshahr. At the same time, the very limited role of pumps reduces the creation of workshops for repair and, possibly, manufacture. As will be seen, however, such trends have not prevented a substantial increase in agricultural wages which have been boosted by the increase in production.

How they live

Since ancient times, as Burton Stein explained it, in the advanced areas of the Coromandel coast there has been 'a basic cooperation between the peasant cultivators and the Brahmins,' [11] a pattern which we find in Manchala, Guntur district, but which had been lacking in the much less developed part of Thanjavur which came under canal irrigation only after 1935. In Kila Ulur, the dominant caste is the Kallar, followed by the Ambalakars and then the Paraiyars, a Scheduled Caste.

The fact that there were no 'twice born' castes did not prevent caste barriers from being very strict, a situation which began to change after Independence. In 1964, when questioning the Paraiyars about possible changes and progress in their conditions of living, they replied that their material life had not improved substantially, but several caste discriminations had gone. By that time they were no more compelled to walk barefoot through the Kallar settlement. They could bathe and wash in the Kallar tank and have tea in the village tea stall. When travelling on a bus, they could even sit near a Brahmin. [12]

[11] *Op. cit.*, P. 69., although it was not a universal trend.

[12] Such changes are also common in the old delta, but there are villages where

No doubt life was hard, but one did not notice striking abuses or pressure from the Kallars. The Paraiyars had little land and worked, full time or part time, as agricultural labourers. They also made mats in the lean season. They collected fish in May when the surrounding tanks were becoming dry. They were also able to cut some wood in the casuarina forests, find an occasional job involving the repair of canals or the building of local roads.

From 1964 onwards, the situation of the Paraiyars and other small farmers did improve in material terms.

	Daily wages in Rs.	*Daily wages in kg. of paddy for harvest and threshing*
1964	1.00– 1.50 (rarely 2.00)	4.8–5.5
1978	5.00– 6.00	7.2–9.0
1985	7.00–10.00 (sometimes more)	15.0–22.0

	Government procurement price (in rupees) paddy/q.	*Wholesale price (in rupees) paddy/q.*	*Retail price (in rupees) Free market rice/kg.*	*Retail price (in rupees) Government fair price shop*
1964	35.00–40.00	66.00–75.00*	1.0	0.73
1978	85.00	117.00–178.00*	1.70	1.60
1985	140.00–146.00**		3.40–4.50*	2.25

*Seasonal variations

** The Tamil Nadu government pays a subsidy of Rs. 15 per quintal not included here, which was increased to Rs.35 at the end of 1985.

Wages at the time of harvest have clearly improved. As for wages in cash, the increase is more obvious between 1964 and 1978, than between 1978 and 1985, especially if the landless or small farmer cannot buy grain from a fair price shop where it is subsidised and is forced to go to the free market.

The rising production has an obvious impact on wages especially for the *kuruvai*, when the harvest has to be completed in record

such changes nave yet to come about. In December 1985, I was in a village near the national highway, 15 km. from Madurai, where Harijans were still prevented from wearing shoes in the streets of the Nadars and Naikers, from walking with an open umbrella or from wearing their lungi tied up, thus showing their legs!

time for fear of an early north-east monsoon.[13] Landowners thus indirectly contribute to the wage increase by being forced to employ more labour for the harvest.

Another factor enters the picture. Since the 1940s the district has become politicised by an active Communist Party so that there has been some kind of agrarian trade unionism, a rare phenomenon in India. This has periodically led to tensions, sometimes even very bloody incidents (1967 to 1969), which, as a result, have contributed to wage increases. Thus, the share of the harvest left to the workers increased from one-twelfth in the early 1960s to one-tenth (1967) and, since 1980, to one-eighth.

These political factors have played a significant role in the old delta. Although the new delta is much less affected by the Communist Party movement, the wages have followed the same trend.[14]

One must finally underline how difficult it is to assess wages precisely. There are variations from one case to the other. Food may or may not be included. For pulling (a man's work), transplanting (normally done by women) and harvesting (done by both), wages are calculated on the basis of how fast people work, hence again there are some variations. One can also come across different arrangements in the case of workers from outside. In 1985, I met two young men who, coming from another district, were paid 10 kg. of paddy and three meals per day for ploughing, whereas local people would get Rs.7 and one meal, rarely two.

Wages for men and women differ. A man, depending on his strength and ability, can earn Rs.6 to 10 by pulling shoots from the nursery. A woman earns Rs.6 to 7 for transplanting. Wages can also be paid on a contract basis. There is, for example, the case of a group of Kallars who have come from another village for the harvest of *samba*. Two men and their wives, and two boys do the job. In two days they cut 0.6 acres (0.24 ha.). They keep 137 kg. of paddy for themselves (one-eighth of the crop). They also receive one meal per day.

Regarding wages for groundnut cultivation, they amount to around Rs.7 plus two meals. Daily wages for casual work outside

[13] Tamil Nadu benefits from the south-west monsoon (July-September) and from the north-east monsoon (October-December).

[14] For these political problems, see A. Béteille, *Studies in Agrarian Social Structure* (New Delhi, Oxford University Press, 1974), and, K.C. Alexander, *Agrarian Tension in Thanjavur* (Hyderabad, NICD, 1975).

agriculture are even more difficult to measure. It can be Rs.4 or even double that.

In this village, as in many other paddy growing areas, poor people belonging to the low castes recover the grain stored by rats in the bunds surrounding the fields. They are sometimes able to gather several kilograms. [15]

During my visits in 1978 and 1982, I noticed the impact of the trend in wages referred to above. Many landless people or very small landowners were able to improve their conditions of living. The sight of the Paraiyars hamlet was refreshing: better clothed women, lively children. Chinaya (with 0.4 ha. of land) told me: 'now we live in a civilised manner.' More children of poor families attend school. Several Paraiyars have found a job outside and send money home.

In 1985, however, the overall mood in the village was less positive. A very severe drought had considerably reduced the flow of the canals, so that the *samba* crop, soon to be harvested, [16] was in poor shape. Besides, since 1982, production had levelled off, a phenomenon confirmed at the district level as seen later. However, one could not talk of a real decline in standards of living.

Purulai is a Paraiyar. As in 1978, he owns 0.8 ha., all double-cropped with *kuruvai* and *thaladi*. In 1984-85, the lack of water affected the latter crop. This year, (1985-86), *kuruvai* was sown and transplanted late, also for the lack of water, with the result that he harvested 2,740 kg. out of 0.8 ha., instead of 3,420 kg. in 1984 (equivalent to, respectively, 3,420 and 4,275 kg./ha.). In spite of this fall in production, Purulai is not in any trouble. He had borrowed money in order to give his eldest son a good education who, in 1978, was offered a job in Kuwait. Not only did he get it, but his brother also left some years later for Dubai. Only one son and his family stay at home to look after the father. The two other sons send part of their savings home.

P.Nagamutha, Paraiyar, has not done as well. He owns 0.75 acres (0.3 ha.) where *thaladi* follows *kuruvai*. He is now aging so that when he works as a labourer, he cannot earn more than Rs.6 a day for pulling paddy out of the nursery. (As seen above, a

[15] A typical source of income which does not enter the collection of data on income! I learnt of this practice through M.N. Srinivas.

[16] I was in Thanjavur by the middle of December.

strong man can earn Rs.10.) His wife also works at transplanting and harvesting. He has two boys not yet married, and a girl. Nagamutha has two sets of clothes. It is not easy to make both ends meet. After harvesting barely 910 kg. of *kuruvai* in 1985, he had to sell most of it to meet various expenditures. He was worried for his *thaladi* because of the drought. Like all other very small farmers, he works hard and uses new seeds, chemical fertilisers and some pesticides. Yet, if water fails in the canals, the other inputs can have only a limited impact.[17] Besides this, he also sells some coconuts from his trees and some eggs and chickens.

Several Kallars are also small landowners. Here is the case of a man with 0.6 acres (0.24 ha.). Like most other small landowners, he grows two crops on all his land, getting the equivalent of 3,300 kg./ha. per crop. He could get more if he used urea and di-ammonium phosphate. However, he does spray against pests. His wife and he also work as labourers. While he is not really miserable, in the past few years he has not seen much improvement in his life. Progress had been a little more significant in the 1970s.

Tangavelu belongs to one of the five families of Dhobis (washerman). He works for fifteen households, each giving him 40 kg. per year for his services. In addition, he owns 0.4 ha. of land from which he gets 1,710 kg. of paddy (equivalent to 4,275 kg./ha.) followed by groundnut. He also has a small bullock. Young and well built, he wears short trousers and a shirt. His wife has five saris. They have a radio, but they have to repay a debt of Rs.7,000 at 24 per cent interest. Tangavelu has cleverly taken advantage of irrigated groundnut and his yield of paddy is high which implies adequate doses of chemical fertilisers. On the other hand, the amount of paddy which he receives for washing clothes has remained the same for the past twenty years, a typical example of the weak bargaining power of the Dhobis *vis à vis* the agricultural labourers.

In 1978 itself, it was striking to observe how even farmers with holdings below 1 ha. could manage to sell some paddy (thanks to rising yields) after covering their home requirements. R.Tolasi, a Kallar, lives out of 0.8 ha., with his wife and three daughters, not yet married. Near his house, there is a pair of bullocks and three

[17] As explained previously, new seeds-cum-chemical fertilisers require more water than traditional varieties without chemical fertilisers.

cows. He first grows *kuruvai* paddy getting the equivalent of 4,275 kg./ha. He sells one-fifth of the paddy, i.e., 860 kg. Then he grows groundnut, all of which is sold. His entire family helps in the fields, but he needs extra labour for transplanting and harvesting paddy.

Palamivelu is 23 years old. His parents have died, so that with his younger brother, he owns 2 ha. of land. In addition to paddy, he has started growing sugarcane on 0.4 ha. Obviously, the two brothers are quite well off.

Duraj Raja Peryar, a Kallar, owns the same amount of land. He lives with his wife, two sons and one daughter, while the eldest son is at college in Thanjavur. Having installed a well-cum-pumpset, he now starts his nursery of *kuruvai* early.[18] This ensures a good yield of 4,275 kg./ha. of paddy, followed by *thaladi* at 2,850 kg./ha. After that he irrigates some fields in the dry season where he grows groundnut. He spares 1,140 kg. of paddy for sale, followed by groundnut. He can afford three cows (which together yield around 5 litres per day), and two pairs of bullocks which he rents from time to time. He owns a radio and a bicycle. On the whole, he is quite well off.

Finally, we visit one of the big landowners, Krishnamurti, a Kallar. Owing to his 6 ha., he has managed to buy a tractor with a government loan, for the price of Rs.150,000. He devotes 2 ha. to *kuruvai* followed by 0.8 ha. of *thaladi*, the former giving 4,275 kg./ha., the latter only 3,000 kg./ha. for lack of proper irrigation. On 1.2 ha. he grows groundnut. Four ha. are devoted to *samba*, which also yields 3,000 kg./ha. However, he complains of insufficient irrigation. Last year, his crops suffered from an attack of pests. In addition to his sales of paddy and groundnut, Krishnamurti rents his tractor (Rs.120 for two ploughings of 0.4 ha.) and also uses it for transport with his trolley. For a distance of 30 km. he charges Rs.140. His driver receives Rs.225 per month.

We held several meetings with groups of Kallar landowners. No doubt, their life has improved in the past twenty years. They eat better, they are replacing earthen pots with aluminium vessels. Men and women, they say, use better soap and now even make use

[18] He starts at the beginning of June. After twenty-five days paddy is transplanted. By the middle of September it can be harvested, well in time before the north-east monsoon which can damage paddy if it is still in the fields, as may happen when farmers have to wait for the arrival of water in the canals.

of talcum powder! They also have some toddy and arrack shops. Bicycles and radios are more common, three rich men have each bought a scooter or a motorcycle.

As seen above, the very small landowners and agricultural labourers have not been left out of this gradual process of change. However, at all levels, people don't fail to insist that the rising cost of living is eating away part of their income.

When calculating the cost-benefit ratio for 0.4 ha. of a good *kuruvai* crop of 2,000 kg.(5,000 kg./ha.), the gross profit is around Rs.3,000, the net profit Rs.1,500, which rises to nearly 2,000 when the cultivator does not rent a tractor but uses his own bullocks and when he reduces the number of labourers employed for transplanting and harvesting.[19]

All farmers interviewed agree that the growing constraint of poor water management undermines the boost of yields which, with better irrigation, is perfectly possible.

I then raised the question of family planning. Although it is more successful here than in the Ganges basin, much more could be done. 'Are you not convinced of that', I argued, 'since in the next generation, holdings will become again divided and smaller?' 'Yes we understand your point of view,' replied several landowners, 'but we do not like to go to the camps where you are sterilised, often under improper conditions,' a complaint often recorded in villages, which confirms the need to improve the family planning programme, a point, unfortunately, more often merely stated than achieved!

Finally, we came to marriage problems. In the 1960s a Kallar farmer with about 2 ha. of land spent Rs.1,000 for a daughter's marriage. Today the expenditure is ten times greater. Poor Kallars may have to spend upto Rs.6,000.

Thanjavur district

Although the rate of increase of population is relatively low, the average density per sq.km. is becoming heavy : 394 in 1971, 490 in 1981. Orathanad taluk (population 4.05 million, area 8,280 sq.km.)

[19] One cannot be accurate since labour charges vary according to the size of the owner's family. Yet, even holdings of 0.5 to 1 ha. require at least some paid labour unless there is mutual aid between families.

where Kila Ulur is located, only reached 366 in 1981, as a consequence of a lighter population (compared to the old delta)[20] until the advent of canal irrigation. Although we are in an advanced district, the rate of urbanisation is not so high : 23 per cent, slightly below the all India average and much below Tamil Nadu (33 per cent).

As in so many other parts of the country, there is no more land to reclaim. The area devoted to *samba* has reduced over the past twenty years in favour of *kuruvai* and, possibly, *thaladi*. The total gross area under paddy has not changed much while there has been some increase in multi-cropping for other crops.

Land use (ha.)

	1984-85	
Kuruvai	194,000	
Thaladi	168,000	
Samba	266.000	
	1974-75	*1984-85*
Sugarcane	6,065	8,767
Groundnut	38,564	44,450
Gingelly (oil seed)	5,627	14,775
Green and black gram (pulses)	106,000	172,000
Cotton	negligible	7,240

These data confirm the slight decline in the dependence on paddy in favour of other crops.[21]

Net sown area	510,000–540,000 ha.
Gross sown area	740,000–800,000 (including multi-cropping)

The total irrigated area is equal to 75-85 per cent of the gross and net sown area. Canal irrigation plays a dominant role, followed by tubewells and filter points (30,000 ha.) and tanks (16,000 ha.).

While the district is one of the most advanced in the whole of India, with reference to irrigation, the old rice bowl is facing rising difficulties. In a traditional system of agriculture, requiring less water than fields sown with high yield varieties of paddy and fair

[20] It is not uncommon to come across villages with 700, 800, sometimes 1,000 or more per sq.km.

[21] Data supplied by the district administration.

amounts of chemical fertilisers, even low irrigation could enable farmers to harvest around 2,000 kg./ha. of paddy per crop. Now, along with the new inputs, adequate irrigation boosts yield upto 4,000-5,000 kg./ha., whereas yields fall to 2,000-2,500 in a bad year. In other words, variations in the weather have a greater impact on production now than before.

Rice production (million t.)

1974-75	1.07	1980-81	1.36
1975-76	1.28	1981-82	1.47
1976-77	0.75	1982-83	1.82
1977-78	1.00	1983-84	1.67
1978-79	1.23	1984-85	1.48
1979-80	1.23		

	Consumption of chemical fertilisers (in tonnes of nutrients)	*Of pesticides (in tonnes)*
1974-75	47,900	5,305
1975-76	66,730	17,037
1976-77	59,170	—
1977-78	68,570	—
1978-79	74,000	—
1979-80	76,460	—
1980-81	77,740	—
1981-82	83,720	19,240
1982-83	78,180	—
1983-84	76,590	17,290
1984-85	90,104	15,380

How should one analyse all these data? Thanjavur has moved forward with the introduction of new seeds, fertilisers and pesticides; the irrigated area has increased (only moderately) with tubewells and filter points; but the quality of canal irrigation has not improved. Since canals still play the dominant role in irrigation, this means that rains remain a crucial factor, unlike what has been observed in Bulandshahr where tubewell irrigation is dominant, and, at the same time, relatively safer in spite of a poor power supply. If water is lacking in Thanjavur, farmers will rightly reduce the use of chemical fertilisers (1976-77 or 1982-83). If there are severe floods, the benefits of fertilisers will be lost as in 1983-84.

The rain factor is complex, because it does not only involve the total amount of rainfall but its timely dispersal. Further, though more rarely, the delta may suffer from floods. [22] The first problem is the early opening of the canals, provided the north-west monsoon starts by early June. In 1984-85, for instance, the canals began releasing water by the middle of June, so the *kuruvai* could start at the best time. The harvest took place early enough — by the middle of September — before any risk of an early north-eastern monsoon could have damaged the harvest. In 1974-75, on the other hand, water was released only around 20 July, postponing the harvest towards mid-October or later, by which time the north-eastern monsoon had set in. Between these extremes, the more common time for opening the canals falls at the end of June/ early July, so that the calendar for a good *kuruvai* is pretty tight.

Other troubles may occur when after a timely start of the monsoon, there is a long spell of drought. Particularly tragic was the situation during our visit in 1985. Because of heavy rains in November, the *samba* and *thaladi* crops were in excellent condition, except where there had been floods. But, by the middle of December, due to drought, the Mettur reservoir could supply only one-third the required amount of water. Instead of arriving in December, some heavy showers occurred only by the middle of January, i.e., too late to improve the crops.[23]

How can the situation be altered? There is no more irrigation potential to tap in the interior of the peninsula [24] so that farmers have to manage with the available surface water. One remedy would be to improve the maintenance and operations of the canal systems. The defect noted in 1964 had not changed much during my following visits. Desilting of canals, re-modelling, better distribution, new structures, realignments and lining, if not the main canals, an extremely expensive task, but of the field channels. A combination of these factors could bring down the amount of water losses which may range from 40-50 per cent. A better distribution whereby farmers at the head of the canals use

[22] Average rainfall is 1,200 mm. which can drop to 620 mm. as in 1980-81 and 750 in 1982-83 or rise to 1,846 as in 1983-84, producing severe floods. In 1985, 40,000 ha. of *kuruvai* and *samba* were destroyed by excess rains in November.

[23] *The Hindu,* 17 December, 1985 and *Indian Express,* 18 December 1985.

[24] The catchment areas of the Cauvery are in Karnataka, which creates the additional problems about sharing of water between both states.

less water so that farmers at the lower end are not neglected, would also help.

Then comes the system of irrigation proper. Instead of having field channels or water courses reaching, if not one single field, at least a smaller area, farmers practice the extremely wasteful system of field to field irrigation.

All these problems, in existence for many years, are to be found in many other districts as well. Yet, remedies are slow in coming. They would imply drastic changes in planning and in the allocation of resources, in order to devote more money to the improvement of existing canals instead of investing heavily in new projects, the construction period for which is practically always behind schedule.[25] Money is not the sole problem. There is no doubt that the irrigation or public works department could operate in a more efficient manner, with better discipline, less leakages, if not corruption, again an observation not confined to Thanjavur.

How far can one compensate for the insecure supply of the canals by tapping ground water through deep tubewells or filter points?[26] The cost of using ground water to supplement surface water may discourage some farmers because they may need it only when the rains fail or are erratic, rather than every year. Second, tubewells would be crucial for a good crop in the dry season after the canals are closed (end of January/beginning of February). From March to May there are frequent power cuts, so that farmers would have to use diesel engines which are more costly to buy and operate than are electric motors.

One must also consider the consolidation of holdings. So far nothing has been done in Tamil Nadu. Would such a reform induce farmers to expand ground water irrigation as observed in UP? It is difficult to say because some other districts like North Arcot have known a spectacular development of tubewells and filter points in spite of the fact that there has been no consolidation of holdings. Finally, not all parts of Thanjavur are fit for tubewells. Near the coast, ground water is often brackish making it unsuitable for irrigation.

[25] I will return to this matter in Chapter 14 of this book.

[26] Tubewells of 7-10 H.P.; water table at 130-200 m., total cost Rs.100,000; command area 4 ha. of paddy and 4 ha. of other crops requiring less water. Filter point of 5 H.P.; water table 25 m.; cost Rs.15,00; command area about 1 ha.

Although a massive programme of canal rehabilitation has not yet started, a project sponsored by the World Bank is a first step in the right direction. The objective of the 'On Farm Development Project' is to construct additional channels to ensure water to blocks of 10 ha. each and drain the excess water during heavy rains. [27] Structures for water control and proper distribution of water will be constructed so that 460,000 ha. can be covered in Thanjavur. A first scheme began in 1968-69. By 1984-85, 78,000 ha. had benefited from the project at a cost of Rs.39.3 million, i.e., Rs.5,040 per ha. [28] On the basis of preliminary results, the incremental benefit in crop yields is in the range of 10-20 per cent.

There are other advantages of a more careful use of water. It is recommended that the fields be dried for a few days at the tilling stage (twenty to twenty-five days after transplanting) because more stalks appear if the soil is aerated. After flowering, when the grain appears, water is crucial, but later on, twenty to twenty-five days before the harvest, it is better to drain the field. These practices can increase the yield by 5-7 quintals per hectare.

To sum up, one can say that, *a)* after having been a factor of growth for two thousand years and, more recently, of rising yields, with the addition of new inputs, irrigation is now the major obstacle to further progress, *b)* year to year variations seem to increase compared to earlier times, when new inputs were hardly available.[29]

Paddy yields for the district (kg. per hectare)

	1964-65	1975-76	1984-85
Kuruvai	2,640	3,640	4,600
Thaladi	2,260	2,715	3,100
Samba	2,700	3,040	3,000
Combined	2,645	3,225	3,725

All three years have enjoyed good weather conditions. If we compare the average combined yield of 2,250 kg. per hectare in 1959-60 with that for 1984-85, the total increase amounts to 66 per cent

[27] Other major drainage works are also needed.

[28] Part of the cost is borne by the farmers through loans repayable in five years.

[29] Variations would have been around 15 per cent then, whereas they can be greater nowadays, as seen above.

in twenty-five years, which is not so outstanding. Besides, if we compare the total production from 1975-76 to 1984-85, the increase is seen to fall to 16 per cent which is below the population growth.

Would it be possible to push yields to the level of 5,000 kg./ha. per crop ? This is not certain, even when taking into account the improvements in water management referred to above, because all weaknesses cannot be eliminated. If this can nevertheless be achieved, the increase would be 34 per cent, not so high keeping in view the future population increase.

Other alternatives are being discussed and studied. Paddy consumes much more water than most other crops, so why not alter the cropping pattern ? Such a trend is already noticeable (see data above), but how far it can go remains to be seen. Other district cadres are thinking of changing the calendar of water release, provided there are more tubewells or filter points. They would irrigate *kuruvai* which could be followed by cotton or groundnut. The canals would be opened from November to March, thus releasing more water in five to six months, instead of seven to eight months. As referred to earlier the greater emphasis on tubewells may, however, be hindered by the lack of electricity.

Land tenure and employment patterns

The new delta not only has a lower density per sq. km. (366 in Orathanad block) than the old delta (average density of the whole district, 490 sq. km.), it also has a higher ratio of cultivators to agricultural labourers.

	Orathanad taluk	*Thanjavur district*
Number of cultivators	39,000	350,000
Number of agricultural labourers	32,000	650,000

(1981 Census, round figures)

The district data give us an idea of the acute division of holdings which has taken place.

		1981 Number of holdings	*Area*
Below	0.5 ha	245,315	52,144 ha
	0.5 - 1	116,450	86,183
	1 - 2	100,757	148,871
	2 - 3	38,367	95,616
	3 - 5	24,198	92,429
	5 - 10	11.258	75,245
Above	10 ha.	1,892	36,799
	Total	538,237	587,287[30]

Not having worked in the old delta, I cannot say much about the standard of living as compared to the area of Kila Ulur. Wages at the time of harvest have risen from one-twelfth to one-eighth of the crop. Cash wages imply some possible improvement in real terms, but this would need confirmation. About the village of Palakurichi, G.S.Guhan states that 'in the long span of more than four decades...real earnings of agricultural labour have not gone up even if they have not decreased.'[31] On the other hand, extreme caste barriers have disappeared and education has progressed, even among the Harijans. Several of their huts are electrified and a third of them now have their own radio and bicycle.

Taking into account the very large number of landless labourers and small farmers, one cannot but be worried about the future even if some improvement has also occurred in the old delta. Until now, the intensity of cropping, the high rate of irrigation and the fact that paddy takes up around 75 per cent of the land, contributes to a high demand for labour, including female labour. The demand is such that in the peak season, even workers from neighbouring districts are needed. The introduction of new varieties of seeds and the rising yields have also contributed to more employment, but there are limits to this process.

As noted earlier, decisive improvements in water management are far from simple and yet they are of utmost importance, not only in order to push existing crop yields further, but also to allow

[30] Data supplied by the district headquarters. According to the public works department there are 504,698 holdings and 656,857 ha. The break up of holdings is roughly similar.

[31] See the fine study of S. Guhan, *Palakurichi, a Resurvey,* Working paper no 42, Madras Institute of Development Studies, November 1983. This is one of the villages studied by G. Slater in 1916 and re-surveyed in 1937.

for more treble-cropping in the dry season, such as black and green gram or other crops.

The infrastructure in terms of good roads and the quasi-total electrification of villages could also enhance the diversification process of the economy in order to create more employment outside agriculture, especially in small industries, but, again, provided the supply of electricity improves substantially. It is indeed surprising to see that the rate of industrialisation is fairly low : Thanjavur accounts for only 2 per cent of the factories in the whole state and the active population engaged in household industry, processing, repairs and manufacturing amounts to 30 per cent of the active population versus 70 per cent in agriculture. This goes along with the relatively low rate of urbanisation : 23 per cent as against 33 per cent for the state as a whole. It should be possible to expand the secondary and tertiary sectors more rapidly, especially in the small and more important towns.

One should also consider fishing and aquaculture. In 1978, the collector of Thanjavur had started some shrimp ponds (in brackish water) near the sea. They looked very promising but when I enquired in 1982 and 1985, the projects had proved unsuccessful, which, certainly does not mean that one should give up the idea altogether.

On the other hand, in the past years, sea fishing has increased. The fishermen use trawlers owned by traders, who then sell the fish to pedlars involved in the retail trade of dry fish. Most of them are marginal farmers and landless labourers belonging to low castes. The consumers also belong mostly to very low castes.[32]

Like all other districts, Thanjavur is increasing the number of special programmes for the poor. The DRDA (District Rural Development Agency) promotes various schemes : water tanks and taps, milch cattle supply to the poor, etc. As seen in Kila Ulur the cattle-cum-milk supply scheme which looked encouraging in 1978, has broken down. The drinking water system also seems to be well behind schedule. The DRDA would like to supply more crossbred cows (supplying 6-10 l. per day) to replace local cows (1 - 2 l.), but they don't get enough cattle. They also provide (on the basis of a loan-cum-subsidy), sprayers for pesticides, and bicycles.

[32] H.G. Bohle, Small Trade as a Lower Circuit in South India's Rural Economy, paper presented at Heidelberg, 9th European Conference on Modern South Asian Studies, 1986.

The future of the old rice bowl, which has also been the cradle of Tamil culture, is not so bright and it shows how complex the whole process of development has become. Yet, there is still room for manoeuver in agriculture, in industry and in services. Another asset is the administration. The collector or district magistrate is usually selected with great care because of the economic importance of the district and of the possible risk of social tensions. The collectors, whom I met during my different visits, were IAS officers of a high calibre. They were competent and active, officers dedicated to development and they knew their villages intimately. The joint director for agriculture and his team have always been more efficient and involved than some of their colleagues elsewhere. More than ever will such people be needed. At the same time, they would be no less welcome in the public works department, where efficient officers could achieve a great deal.

Chapter Six

Another Advanced Area

As in 1978, we drove by car from Orissa to Andhra along the coast. What a contrast between the villages and towns of Orissa—certainly not dormant but not bursting with traffic, trade and industries — and the buoyant economy of east and west Godavari, Krishna and Guntur districts. These contrasts are even more prominent now. Superb plantations of coconut and banana trees, sugarcane and fine paddy fields dominate. The traffic is heavy not only with trucks and buses but also — though not so frequently — with a fair number of cars belonging to rich farmers or businessmen. There are also a growing number of motorcycles and motor-rickshaws. Rice mills, other factories, cinema halls and modern air conditioned restaurants add to the air of affluence.

Obviously, the links between cities and villages are too strong for this expansion to be confined to towns alone. To assess the changes we once again took the road from Guntur to Manchala in Tenali *taluk*.[1] After crossing the wide Buckingham Canal, we reached Manchala at the time of the harvest in November 1985.

As in other villages, the geographical and cultivated areas have remained the same for many decades : 313 ha. and 266 ha., respectively. Paddy dominates in most fields during the monsoon. It is later followed by pulses (black gram) using the moisture left in the soil. Around the houses stand coconut, bamboo and other trees.

The population has risen from 1,462 in 1961 to 1,579 in 1981, i.e., an increase of only 8 per cent. This is very surprising, *even* if we bear in mind that out of the twenty Brahmin families living here in 1967 (my first visit), many of them have left the village to settle down in towns. Several Vaishya families have also moved. From census data, it appears that several poorer people have also

[1] Tenali *taluk* has a very high urban population: 44 per cent of the total. The district itself has 28 percent of the total (1981 Census). The town of Guntur had 367,000 inhabitants in 1981, Tenali 119,000.

left.[2] Family planning has had some influence. Even in 1978, as many as 100 women had been sterilised, a trend which continues today. Despite the low population figures, the density per sq.km. has nevertheless increased, reaching 504.

According to the 1981 Census, there are 377 households. The Kapus are the dominant caste, owning the largest part of the land. There are only a few families belonging to Scheduled Castes but there are 156 Yanadi, a Scheduled Tribe, who settled in the region long ago.[3]

I could not get precise information on land holdings. The 1981 Census mentions 377 households or families whereas the 1981 Agricultural Census indicates 511 agricultural holdings. This discrepancy cannot be explained merely by the fact that large landholders claim to have divided their land in order to avoid the ceiling (4 to 6.4 ha. of irrigated land).[4] The records also mention that 447 people are landless.

In spite of these unreliable data, it is clear that the break up of landholdings is not exceptional. There are an increasing number of small to very small holdings. When leaving Manchala, the Brahmins sold their land, but in parts, so that it did not give the impression that the number of large landholders was increasing. For instance, my friend S.Sivananda, a Brahmin, sold 4.8 ha. to more than twenty Kapus and his cousin did the same.

Growth trends

In the late 19th century, agricultural growth in Guntur district relied mainly on the expansion of rice and tobacco due to the new canal irrigation system. Manchala lies in the area where rice is the main crop. Traditional techniques had more or less been saturated towards the early 1960s. Kapus and Yanadis are good farmers; fields are properly levelled; nurseries are raised with

[2] Between 1971 and 1981, the number of Scheduled Tribe people seems to have fallen from 183 to 156. Again, urbanisation alone cannot explain such a low rate of population increase.

[3] I do not know if it was during the last century or earlier. Originally, Yanadis lived in the forests in the interior.

[4] These discrepancies are puzzling, both as to the total number of people and to the number of holdings. In 1967, according to a rough estimate, there were 174 holdings, out of which 143 were below 1.6 ha., in 1978 a total of 240.

care ; transplanting is done in a brisk and efficient manner like the other operations: weeding, harvesting and threshing.

Farmers grow one crop of paddy, because canals release water from July to the end of the year. One week before the harvest of paddy, pulses are sown broadcast. It is a low cost crop as it does not need ploughing. Besides, the pulses enrich the soil by fixing nitrogen from the air.

Yields of paddy (kg./ha.)

1955-1960	2,250 - 2,500
1967	2,800 - 3,500
1978	3,750 - 4,125
1985	4,000 - 4,700

While there has been no decisive qualitative change in irrigation, which covers practically all fields, chemical fertilisers appeared around 1960. Improved seeds also contributed to higher yields, but not to the same extent as the high yield varieties which respond well to higher doses of chemical fertilisers. The early new seeds, which we saw in 1967 were not well adapted to local conditions. Even during our second visit, new seeds continued to run into problems, whereas now, recent varieties give better results. These teething problems which one did not encounter with wheat are due to several factors : paddy is cultivated under a wide range of micro-climates which may explain the success of the same variety in one place and not elsewhere. Second, monsoon crops are more exposed to pests than irrigated crops in the dry season. Third, in such technical transfers one cannot ignore the factor of luck: good luck in the case of Mexican wheat and bad luck as with the failure of paddy varieties imported from Taiwan in the late 1960s.

Pest control, practically unknown thirty years ago, is quite widespread now. Farmers are well aware of its value and advantages and are particular about the types of products they want.[5]

There has been no change in pulses (black gram) since my first visit in 1967. Yields vary between 500 and 750 kg./ha. but the prices have risen substantially, a trend common to the whole of

[5] In 1978, there were demonstrations in Guntur because the administration was not supplying the brand of products popular with the farmers. Now there are lots of complaints of adulteration of products by small dealers!

India, considering the rather stagnant production *vis á vis* the rising demand due to the rising population : the retail price of black gram was Rs.1.40-1.75 per kg. in 1967 and Rs.6 in 1985

Ploughing with bullocks had practically disappeared by 1978. Today there are six tractors often rented by their owners to other cultivators.

During my visit in 1967, I was impressed by the diversity of activities, especially among the poor. In addition to their work in the paddy fields, Yanadis made bamboo traps to catch field rats. They sold these traps and hunted rats for landowners. Some Yanadis were engaged in petty trade, others would collect hair for sale in the bazars to make wigs. Yanadis would also catch birds and fish and were occasionally involved in the construction of houses or the repair of canals.

Electricity reached Manchala in the late 1960s. This paved the way for the construction of a small rice mill. Trade and transportation between Manchala and Chebrol, the nearest major bazar, and Guntur, has been increasing since the late 1960s.

One now notices a fair number of new houses. The few rich landowners live in fairly modern houses of brick and sometimes, concrete. The Yanadis (see later) have benefited from a government housing programme for the poor. However, we don't come across the same construction boom as in Bulandshahr district.

There are a few more shops selling groceries, grain, soap, onion, cigarettes and a few tea stalls near the primary school.[6]

Living conditions

In 1967 one hardly found the type of acute poverty observed at the same time in north Bihar. My observations fit quite well within the overall description of coastal Andhra by Daniel Thorner, referred to in the previous chapter.

People had made proper use of the canal infrastructure, as well as of the general development of the district. Brahmin landowners did not plough their land themselves, but they supervised the work of their Kapu or Yanadi labourers very carefully. Kapus proved good agriculturists. The Yanadis were clever at finding all

[6] Literacy is quite high: 347 males, 183 females, a total of 530 out of a population of 1,579.

sorts of petty jobs in addition to their work in the paddy fields and later, for harvesting pulses. Well built, quick and hard working, they were a fine sight to see in late July, pulling the shoots from the paddy nurseries, while dozens of well dressed women transplanted paddy in properly levelled and watered fields. Later, in November 1978 and November 1985, I was impressed by the way paddy was harvested and threshed. Many landless and small Kapu cultivators were and remain no less clever.

Although the global process of rural development has had less of an impact here than in Mirpur, the overall standard of living remains relatively high. Real wages and employment opportunities have increased. This may be due in part to a lower rate of population increase, the impact of higher urbanisation, lower local density per sq.km and possibly other factors as well.[7]

	Agricultural wages for paddy cultivation (in Rs.)
1967	3.00 (Sometimes 4.00)
1978	5.00 - 6.00 (up to 8.00 for harvest)
1985	10.00 - 12.00 (sometimes up to 15.00-20.00)

Wages are now paid in cash even for the harvest. They can be arranged on an individual or a contract basis with teams of men and women. Men till the land and pull out the shoots. Women do the transplanting and weeding (sometimes men are also involved). Both men and women take part in harvesting and threshing. For weeding (not included in the above data), wages have increased from Rs. 1.50 (1967) to Rs.3 (1978) and Rs. 5 (1985). Normally no meal is included, except in the case of threshing. The paddy fields keep people busy for about three months. Then they spend a few weeks harvesting pulses. Later, they are involved with the repair of bunds and canals.

	Retail price of rice (Rs./q.)	*Purchasing price of paddy*
1967	100.00	43.00 to 55.00
1978	140.00 to 170.00 (high quality up to 200.00)	85.00* to 95.00
1985	240.00-275.00* to 300.00-400.00	140.00* to 160.00
	*fair price shop	*procurement price

[7] All these comments are tentative. A more detailed study would be needed to confirm them.

The farmers' selling price for black gram has risen from Rs.250-300 per quintal in 1978 to Rs.400-500 in 1985, depending on the season.

One can conclude that on the whole wages have increased a little faster than the price of rice. It is, however, difficult to assess whether or not employment opportunities have also increased. Considering the migrations outside the village, which include poor people, this seems plausible.

One must also take into consideration the social schemes introduced by N.T. Rama Rao after his party, the Telugu Desam, won the 1983 elections. People earning less than Rs.600 per year receive a green card which entitles them to buy 25 kg. of rice at Rs.2 per kg.per month for a household of five, to purchase a *dhoti* for Rs.25, and to obtain some special health facilities. There is one government rice shop in Manchala which seems to operate quite efficiently for that purpose.

In 1978, the Yanadis' wages had slightly overtaken the overall index of the cost of living. More important, they had been allotted a plot of village land (0.4 ha.) on the other side of the road near the school, to build their houses. To be the owners of their homesteads and of the land was quite a change compared to their previous situation when they lived on plots belonging to Brahmins or Kapus who could subject them to various injustices like unpaid services. 'Now we stand on our feet', said a Yanadi. Life was not too precarious and one came across some semi-durable goods in their possession: bicycles, transistors, etc.

The cyclone of 1977 had damaged some of their delicately built huts. Government relief operated rather smoothly, so that each family received Rs.150, one sari per woman and rice for one month.

In 1985, the Yanadis improved their living conditions further and benefited from the new government housing scheme for the poor. Small double bricked houses with a tiled roof have been built to withstand cyclones not uncommon in this region. Each owner gets a subsidy of Rs.4,000 and a loan of Rs.3,000 repayable in ten years at 4 per cent interest. The Yanadis got saplings of coconut trees at Rs.7 and of eucalyptus at Rs.0.10 per piece. The houses were well kept with a mini kitchen garden, flowers and some vegetables. Some banana trees grew near a few houses. One man had started a small shop in a hut, where he sold vegetables, and cigarettes. Inside the houses there were a number of brass pots and plates, some plastic goods, some beds and a coffer with a few sets of clothes. Around the houses ran a few chickens, and one of the Yanadis had recently bought a pig for Rs.300. The drinking water well is on the other side of the road and like the rest of the village before them, electricity was soon to reach them as well.

I must confess that it was remarkable to move around, to talk to cheerful people, people fully aware of such changes in their lives. On the whole, they receive better wages for paddy cultivation. Women remain active and lively as ever.

While they have given up catching birds because it is no longer economical, they continue to make traps and catch rats, thus earning Rs.0.30 to 0.50 per rat against Rs.0.25-0.30 in 1978. For four months they catch fish in the canals, earning Rs.10 per day of catch. Some odd jobs — canal and road repair, some service to the Kapus — bring in further income.

Chalainaya lives with his wife. He had three daughters but one of them died. The other two are at school. After his wife became pregnant again, Chalainaya decided to get himself sterilised. Around the house are two coconut trees which will soon produce nuts for sale. One buffalo supplies 1.5 l. a day which is consumed at home. He has another smaller one, as well as three chickens.

Chalainaya earns from Rs.8 to Rs.12 doing spade work (after ploughing) to prepare paddy fields, then he pulls shoots from the nursery which his wife transplants. Later, he is busy with weeding, harvesting and threshing. Some months later they harvest pulses. In all, agriculture keeps him busy for five months and his wife somewhat less. Chalainaya has several sets of clothes like his wife and he even owns a cycle.

A little further lives P. Palaya, an old man who started as a worker on a rubber plantation in Penang. He still remembers those 'good days' but after returning to India in the 1950s, he has managed to make a fairly good living. His wife is still alive. Their daughter is married in another village. His son, now 40 years old, has two children and lives with them. Their sources of income are similar to those of most other Yanadis: work in paddy fields, fishing and drying small fish, catching rats, along with some other jobs, including transportation to Chebrol. I asked Palaya if they were still 'plundering' the paddy stored by rats in the bunds around the fields, a widespread practice in India. He said no, because it was too much work to sort out the grain from the mud with which it is mixed, an indirect indication of a rising standard of living.

Let us move to the main village. Most of the Brahmins who left in the late 1970s and more recently, sold their land as mentioned earlier. The Kapus fully dominate the scene, a typical example of the changing socio-political, class/caste relations, as seen in many other parts of India as well.[8]

Babla Munta Suryanarain, a Kapu, lives with his wife and two children. He works hard on his land (0.4 ha.). In a good year, with neither excess rain nor lack of canal water, he harvests 1,875 kg. (the equivalent of 4,867 kg.per ha.), but this year (1985-86) he expects only 1,350 (3,375 kg./ha.) He needs about 75 kg. of paddy (50 kg. of rice) per month for his family so that he is able to sell 9-10 quintals of paddy. After this he grows 4 quintals of black gram (pulses) most of which is sold for Rs.400-500 per quintal. Babla and his wife also work as labourers so that they can manage to buy the chemical fertilisers and pesticides needed for paddy and also to be able to have enough clothes. Babla owns a cycle which he uses to transport various goods to Guntur. For this he earns Rs. 10 a day. The relative affluence of the family is due to two sets of factors : a high yield of paddy and black gram and a small family of only four people.

Among the Kapus it is quite common for a son to receive his share of land from his father when he marries and start out on his own. C.H. Sachanarayan got 0.8 ha. from his father. He lives with his wife and his four small children. Like the other cultivators, he has given up bullocks and rents a tractor. He gets about the same yield as Babla and sells half of his produce which brings him Rs. 3,000.

Cost-benefit for 0.4 ha. (1 acre) and 1,875 kg. of paddy:

Inputs including labour	Rs. 1,000
Gross profit	Rs. 3,000
Net profit	Rs. 2,000

These are the best conditions. If for lack or excess water, the yield falls to 1,350 kg. per acre (3,375 per ha.) with practically the same inputs, the net profit falls to Rs.1,000. Such a difference is indicative once more, of the key role of good water management. Labour

[8] In Thanjavur one notices a similar trend of Brahmins leaving the villages.

charges can also be partly reduced through mutual aid in transplanting and harvesting, something that is not very common amongst the farmers.

Like the other cultivators, Sachanarayan grows black gram after paddy, selling 5 out of 6 q. harvested, at Rs. 500 per q. Two years ago he bought a cycle-rickshaw for transport (Rs.1,600). He transports paddy bags (300 kg) to the rice mill in the village or goes to Chebrol or Guntur. For six months during the lean season, he earns Rs.150 per month in transport. Around the house, four coconut trees were planted one year ago and four chickens provide eggs for local consumption. As a result of his hard work (it is no mean task to cycle with a load of 300 kg.!) Sachanarayan is rather happy. He can even afford to see a film in Chebrol twice a month.

In the house next door lives another Kapu family owing 0.6 ha. of land. The man is aging, one son is at high school the other works in Manchala itself. The mother explained their difficulties to us. Their production is rather low, i.e., 3,125 kg. ha. and the father is now too old to work as a labourer. True they have a buffalo and a cow, also a cycle, but they are not really progressing because of the burden of having married their three daughters for which they spent Rs. 8,000 on each of them, borrowing money at 24 per cent interest.

G. Pullaya, a Kapu, presents us with another success story. While his father was landless, he got a small piece of land from his in-laws at his marriage and, later on, bought another plot from one of the Brahmins. Today he owns 0.4 ha. producing 1,500 kg. (3,750 kg per ha.) of paddy followed by black gram. In addition he leases out 0.8 ha. He bears all the costs, gets half the grain and all the straw valued at Rs.1,000 and all the pulses. In addition, he finds time to do casual agricultural work.

His son is married and the two daughters are still young. Recently Pullaya bought a cart (Rs.1,000) and two bullocks (Rs.3,000) with a loan, at 24 per cent interest. He too is in the transport business earning Rs.20-30 per day. In addition, his buffalo gives 2 l. per day which is consumed at home and he has two chickens as well. On the whole the living conditions of the family have improved. Pullaya has three sets of clothes.

Peddi Setti Aukamma lives much better now out of his 2 ha. of land. He has two buffaloes which together give him 5-6 l. per day,

3 l. of which are sold for Rs.3 per l. He also has a cycle. He is fortunate to have a small family—his wife and two young children.

Let us complete our tour of Manchala with a visit to one of the big landowners, Krishna Prasad. His brother is a doctor in Chebrol. He recently sold 4 acres for the marriage of his daughter so that he is left with 10.8 ha. He can afford a tractor-cum-trolley (Rs.120,000) which he uses himself, for transportation and gives out on rent. His fields, depending on their location, yield between 3,100 and 5,600 kg. per ha.

One can make two concluding observations. Having started with a high level of production, the people of Manchala have continued to do well. As in Thanjavur, it is evident how, with a sharp increase in yields, farmers with 0.5 to 1 ha. of land can lead a decent life, how also, landless labourers can maintain themselves or rise above their miserable conditions. Second, there has been no major qualitative change in the process of growth but an acceleration of the trends already in existence in the late 1960s.

While on the one hand family planning and out-migrations play a positive role in this, one must, as in other villages, raise the issue of rising marriage expenditure. Yanadis used to spend only around Rs. 500 for the marriage of a girl in 1978. This has now risen to Rs. 2,000. Kapu marriage expenditure also tends to rise faster than the cost of living. Small farmers now spend up to Rs.10,000 as against Rs.3,000 in 1978.

Tenali taluk (sub-district) and Guntur district[9]

While the pattern of cropping in Tenali *taluk,* with the total dominance of irrigated paddy followed by pulses in the dry season, has remained unchanged since the beginning of this century and even earlier, several other *taluks* have known considerable change in the past twenty years.

Wide tracts in the interior were only rainfed and, therefore, grew mostly *jowar* and *bajra,* a situation which began to change after 1967 when the main canals of the Nagarjunasagar Project began to release water. They rely on a huge reservoir with a dam on the Krishna in the interior of Andhra. The main right canal

[9] Data supplied at Guntur and taken from Guntur *District Statistical Handbook, 1983-84.* See also the excellent article of V.N. Reddy, Pattern of Commercial Farming in Guntur District, *Economic and Political Weekly,* 21 December, 1985.

will have a final command area in Guntur district of 245,000 ha., half of which is now irrigated. The canals brought many changes in the region. *Jowar* and *bajra* nearly disappeared in favour of paddy and cotton.

Percentage distribution of the main crops in the district

	Paddy	*Jowar* and *Bajra*	Pulses	Cotton	Tobacco
1910-1911	16.5	34.7	6.4	7.3	2.4
1960-1961	22.5	25.9	8.7	0.6	8.2
1979-1980	43.3	6.4	17.2	9.5	1.2

Source: V.N. Reddy, *op. cit.*

Cotton which was not irrigated at the beginning of the century, first fell in the area, to rise again with the help of irrigation. It also tended to replace tobacco. Chillie and groundnut have also lost much ground.

	Net sown area	*Net irrigated area (in hectares)*
1910-11	694,000	128,000
1965-66	720,000	202,000
1983-84	600,000*	333,000

* The fall in sown area is due to the bifurcation of Guntur district in 1970. The eastern *taluks* became Prakasam district.

Canal irrigation covers 315,000 ha. The remaining area depends on tubewells (9,000 ha.), on tanks (3,400 ha.), or, wells and other sources. While many *taluks* have seen a considerable expansion of irrigation, there was not much change in Tenali which now, as for the past several decades, has over 90 per cent of the sown area (44,000 ha.) under irrigation, most of it through canals and only between July and December/January. Tubewells and filter points are confined to 1,830 ha. the whole year.

Production has increased not merely because of the change from rainfed to irrigated crops; the growing use of inputs, like new seeds, chemical fertilisers and pesticides, has played an important role as noticed in Manchala.

District yields of principal crops (kg./ha.)

	Rice	*Jowar*	Tobacco	Cotton (lint)
mid-1950s	1,500-1,600	600-800	700-900	75-95
mid-1960s	1,500-1,650	600-800	600-800	100-120
mid-1970s	1,800-1,900	600-700	850-1,000	350-400
1983-84	2,760	625	1,000	690

If we take rice year by year, we get the following data:

1978-79	1,990	1982-83	2,450
1979-80	1,800	1983-84	2,760
1980-81	2,415	1984-85	2,500
1981-82	2,310	1985-86	2,500

Such yields (2,500 kg./ha. of rice equivalent to 3,700 kg./ha. of paddy) are quite a high average for the whole district. How much further can they be pushed?

As far as the delta is concerned, I heard the same complaints and saw the same shortcomings as I did in 1978. As long as drainage is a problem as in Tenali *taluk* it will be difficult to boost yields much above their present level. High yield dwarf varieties of paddy are particularly vulnerable and cannot but suffer when submersed. Would it not be possible to properly implement the Drainage Act adopted in 1969 by the Andhra government? This act provides for drainage projects which have been implemented very slowly.

Then come similar problems of water distribution and utilisation as observed in Thanjavur. Here again the problems are well identified, but the solutions are slow to follow. No less urgent are better maintenance and better operation of the entire canal system and greater efficiency of the administration in charge of irrigation.

The coastal areas are prone to cyclones between October and December. Not much can be done to protect the crops but, especially after the devastating cyclone of 1977, the government has been encouraging the construction of houses where people can seek refuge in an emergency. Besides, there are also schemes to provide poor people with houses which can withstand cyclones. Radio broadcasts of cyclone warnings have also been improved all along the coast, including Tamil Nadu.

As far as the new canals in the command area of Nagarjunasagar are concerned, I cannot make too many observations for lack of information. Since irrigation depends on a huge reservoir, the vagaries of the monsoon should be felt much less here than in other areas which rely on the diversion of the rivers into canals. One serious problem has been the time needed to complete the whole system, from the main canals down to the fields, and even today there is a gap between the potential created and actual irrigation. On the other hand, one cannot deny the substantial benefits to agriculture. The yield of cotton lint has increased spectacularly. It shows that the controversy about major versus minor irrigation should not be carried too far in favour of the latter. There are cases where large scale reservoirs are fully justified, as in the case of Nagarjunasagar, because of the limited potential of ground water.

Tubewells and filter points (shallow tubewells) cover only around 9,000 ha., a comparatively low area although in the delta the potential of ground-water is quite high. Also, they enable farmers to raise good irrigated crops all year through. Here again I suspect that farmers are reluctant to indulge in such investments because it may be less paying here than in the non-canal areas of Bulandshahr. Canal irrigation is very cheap and tubewells would operate only when the supply of canals fail due to drought. Then the full cost-benefit for a better irrigated crop compared to non-irrigated black gram may not be favourable enough.[10]

The pattern of landholdings does not differ much from Thanjavur. There are about 522,000 holdings for an area of 600,000-650,000 ha. About 183,000 holdings are below 0.5 ha., 136,000 between 0.5 and 1 ha., 110,000 between 1 and 2 ha. and 45,000 between 2 and 3 ha. Wherever most of the land is irrigated, farmers can manage on 0.5 to 1 ha. It is just possible for them to make both ends meet now, but the process of division of holdings is far from over.

In spite of the considerable progress in agriculture, the latter cannot by itself absorb all the increase in the active population. In 1981 the district had 3.4 million inhabitants. The rate of urbanisation has been quite high, rising from 17 per cent in 1961 to 27.6 per cent in 1981. The increase is particularly sharp in Guntur *taluk* : 38.4 to 70 per cent and in Tenali, 23.2 to 44 per cent. Guntur city now has around 400,000 inhabitants.

[10]These explanations are only tentative, more detailed studies would be needed.

Industries, tobacco factories and rice mills have been prominent for several decades, but many new industries have followed in the past twenty years : cotton ginning and textile mills. cotton seed oil extraction mills, wood and paper products, metal industries and machine tools and parts. The increasing motor traffic has led to a fast growth of garages and workshops for the repair of trucks, cars and motor-cycles. Services, trade, cinemas and restaurants are also expanding, not only in Guntur, but in Tenali and other cities as well. Banking is growing equally rapidly. A number of small scale industries and household industries are mushrooming in big villages and towns. Electricity has now reached all the villages. The network of roads, including secondary roads has greatly improved.

In spite of these developments, out of a total active population of 1.5 million in 1981, 70 per cent of the workers were still engaged in agriculture (25 per cent as cultivators, 45 per cent as agricultural labourers). In view of the bifurcation of the district in 1971, one cannot compare data for 1961 and 1981, but it seems that both the actual number of agricultural labourers and their percentage has increased. This gives us an idea of the magnitude of the tasks ahead in order that the process of diversification of the economy be able to absorb a larger percentage of landless labourers.

These facts also highlight the need to accelerate family planning, although Guntur district has not done so badly when compared to many other districts. The decennial rate of population increase fell from 22.23 per cent (1961-71) to 20.48 per cent (1971-81).

On the other hand, we have seen that real wages and employment of agricultural labour have tended to increase,[11] an encouraging sign partly contradicting the trends shown by the census. How accurate are data on employment, especially for the unorganised service sector of petty trade, transport on bicycles and other odd jobs?

One is inclined to think that the actual employment situation outside agriculture may be better than is indicated by statistics. Besides, people classified as agricultural labourers have, as seen in Manchala, other jobs as well which are not measurable by

[11] Data collected in Manchala are representative of the delta, the most heavily populated part of the district.

census data or other statistics. If the problem of employment opportunities is already so acute in advanced districts, one can imagine how serious the situation would be in the more sluggish regions which are equally exposed to heavy population pressure.

Chapter Seven

The End of a Sluggish Agriculture

Having visited the more advanced districts, it is time to shift to the eastern plains, starting with Banaras district (now Varanasi) which has entered a transitional phase and is moving towards faster growth unlike other parts of that wide region where changes remain sluggish.

In the 1930s, Indians and British alike were struck by the contrast between the relatively advanced districts of Punjab and western UP and the acute poverty prevailing in the fifteen districts of eastern UP among which is Varanasi characterised by heavy population pressure on land, poor agriculture, weak infrastructure, low rate of urbanisation (though not Varanasi district itself, being the holy city), sluggish trade, etc. The scene was certainly no better further east towards Bihar.

It is paradoxical that at the end of the 18th century, the situation was roughly the reverse. Agriculture was prosperous, enabling a much larger population than in western UP, trade flourished and so did handicrafts. In fact, even in the heydays of the Mughal Empire, shipments of grain were sent from eastern UP, Bihar and Bengal to the north-west in order to feed the large imperial cities of Agra, Delhi and Lahore. Sir Thomas Roe, the first British envoy to the Mughal court, wrote at the beginning of the 17th century: 'Bengal is feeding India'. The gap in terms of economic development widened after the death of Aurangzeb (1707) when the Empire began to break down. While the eastern plains were not affected much, the north-west suffered from successive and devastating invasions: Nadir Shah and the Persians, the Afghans led by Ahmed Shah Durrani and the Marathas. To these may be added the plunders by the Jats and the proliferation of bandits.

We have seen how crucial irrigation is in these areas. In those days, even in normal times, irrigation systems were not well developed, and besides, the irrigation canals fell into disrepair thus increasing the vulnerability of the area to drought.[1]

As C.A. Bayly writes in a remarkable book: 'Benares was one of the fastest growing cities during the years 1750-90. It became the subcontinent's inland commercial capital ... Yet it was also a city which benefited from the sustained agricultural performance of the rich and stable tracts which surrounded it'.[2] Close links were provided between the village economy and the towns—a phenomenon typical of the global process of rural development found in ancient history—so that the marketing of grain, sugar and oil was possible for distances 'up to fifty or sixty miles'. In those days, several of the present eastern UP districts and wide tracts in south Bihar were quite advanced, whereas, except in pockets around Delhi and Agra, many areas in what are now western UP and Haryana, were stagnant or declining.[3]

Before attempting some general explanations of the changes which occurred in the eastern plains, let us observe the situation in two districts: Varanasi in east UP and Muzaffarpur in north Bihar.

The population trap

After 28 km. along the Varanasi-Jaunpur road, we take a small road, past Mangari a big village which is the headquarters of Pindra block, to drive a few more km. on a fair weather road up to Nilkanth. The flat landscape is broken by mango, *pipal* and bamboo trees and small tanks. The village is scattered with distinct hamlets (caste-wise), a feature typical of a region which has been much less affected by invasions than is the north-west where one comes across compact villages.

There is a false air of affluence when we look at these spacious houses with white-washed walls, half rounded tiled roofs

[1] Old canals were restored by the British and new ones constructed later in the 19th century. See the fine analysis of all these problems by S. Commander, Colonial Rule and Economic Subordinations: The North Indian Economy in the Nineteenth Century, *Etudes Rurales*, January-September, 1983.

[2] C.A. Bayly, *Rulers, Townsmen and Bazaars; North Indian Society in the Age of British Expansion. 1770-1870*, (Cambridge, Cambridge University Press, 1983), p. 104 and the map, p. 76.

[3] *Ibid.*

supported by carved pieces of wood, a vestige of the prosperity of the local gentry of the past. But soon the acute poverty strikes you.

In 1964, the total area was 418 ha., the sown area 362 ha., the balance taken up by the built up area, some orchards, tanks, roads, etc. The population was 2,500 (1961 census) with a density of 600 per sq. km. against 444 in Mirpur, a consequence of an already high population by the end of the 18th century, if not earlier, which remained so in later years. In 1861 Mirpur had a density of 163, while in Varanasi district it was between 200 and 240, i.e., double the density in the 1790s.[4]

Farmers of Nilkanth, as anywhere else reclaimed all available land by the last third of the 19th century. Since then, unlike in Mirpur, few technical changes occurred. The crop pattern remained unaltered, except for the introduction of potato. Rice was and still remains the main crop, cultivated in the *kharif* season due to a higher rainfall than in the north-west (about 1,200 mm. of rain in fifty days, most of it during the monsoon). It covers 180 ha. out of 240 ha. cultivated in the *kharif*. The remaining land bears some millet, pulses and fodder. In the *rabi* 200 ha. are sown: 60 ha. with wheat, 40 with barley, 50 with peas, 14 with potato and 32 with pulses. Sugarcane, a one year crop, is cultivated on 53.6 ha.

Most of the *rabi* crops (except pulses and sugarcane) are irrigated by wells relying on the *mot* system (leather bucket pulled by a pair of bullocks). Unlike farmers in Mirpur, in Nilkanth they have not shifted to the more efficient Persian wheel.[5] In 1964 there were no private tubewells, but two state tubewells had begun irrigating 105 ha.

Yields were fairly low. Chemical fertilisers were hardly used except by a few progressive farmers. Unlike in Mirpur, sugarcane was still of the old variety. There had been no improved seeds of wheat. *Gur* (raw sugar) amounted to 2,000-3,000 kg./ha. (5,000 in Mirpur), wheat to 700-900 kg./ha. (1,200-1,300 in Mirpur). The *kharif* crops relied only on the monsoon which itself varies considerably in total amount, timing and dispersal.[6]

Under such conditions, paddy yields ranged between 1,000 and

[4] Bayly, *op. cit.*, p. 89.

[5] Either the water table was too deep; or nobody took the trouble to try.

[6] Paddy needs so much water that it could not be irrigated by the low capacity *mot* system. It would have needed tubewells.

1,350 kg./ha. (700 and 900 kg./ha. clean rice), considerably below the level (2,000 kg./ha.) reached in Kila Ulur in those days. Paddy was not only vulnerable to drought—due to poor drainage, 40 ha. had to be cultivated broadcast because of the depth of the water.[7]

This land use was supported by an irrigation system limited in area and efficiency during the dry season, and a rather high rainfall (ensuring better crops in the monsoon as compared to the north-west). With densities of 100 or 200 per sq.km., the situation was not too bad, even if there was, no doubt, a sizable gap between the incomes of the local gentry and the agricultural labourers whose life must have been hard. The growing population pressure, the lack of technical innovations (which lead to higher yields), the slow expansion of the secondary and tertiary sectors and the decay of certain handicrafts, together led unavoidably to a growing imbalance between population and resources. As for land tenure, the four landlords (Thakurs) each owned around 12 ha., followed by several medium farmers, and a very large number of small and marginal landowners.[8] The number of landless labourers was no less striking.

Can one confine the explanation of poverty to techno-economic factors? In Bulandshahr, our Jat friends used to say: *'Purbi log dhile hein'* (the easterners are lazy or easy-going). Such a statement is certainly too simple an explanation but it is true that the major landowning castes—Thakurs or Rajputs, Bhumihars and Brahmins — are not backed by a strong agricultural tradition like the Jats, a difference which had struck very experienced British officers like Sir Malcolm Darling and others.[9]

In 1964, people of the higher castes observed the taboo that prevented them from ploughing fields themselves, even if they were small landowners. This lack of drive was obvious. Gradually, the members of the gentry were impoverished. The poor, the

[7] Paddy was of a long stem variety, growing directly in the flooded field. It could not be transplanted because there was often too much water.

[8] The poor condition of land records prevented me from reaching even a rough estimate of landholdings. The division was already acute, judging from the number of plots: 2,083 for 418 ha., versus 613 for 276 ha. in Mirpur before the consolidation of holdings.

[9] Sir Malcolm Darling, *The Punjab Peasant* (London, Oxford University Press, 1947), p. 35 (first edition, 1925).

Chamars (same as the Jativs, Scheduled Castes), continued to survive under very hard conditions which were made worse by semi-feudal relations between them and the farmers of the higher castes. The word *dabao* (pressure) would often enter our discussions with landless people.

There were nevertheless a few exceptions. One of the Thakurs, Kamala Prasad Singh, developed his own land, using chemical fertilisers and supervising his five permanent labourers carefully. He has three brothers who are co-owners of the 15 ha. One worked as a barrister in Varanasi, one as the director of a small factory and the third as an assistant in agriculture at Agra University. While Kamala Prasad led a comfortable life, many farmers with 2 to 6 ha. of land were poorer than their counterparts in Mirpur, or in coastal Andhra. They took little part in work on the fields and paid their labourers badly.

On visiting the hamlet of the Kurmis, one found a more stimulating atmosphere. They were typical of those rising middle castes and classes in villages. Clever, hard working and enterprising, they had behind them solid agricultural traditions, including vegetable gardening, which so struck British administrators in the 19th century.[10] Kurmi cultivators of 2 ha. or less were not so badly off, doing more double-cropping than others, pushing potato. A number of them were able to improve their standard of living, but at the same time they entered into growing clashes with the high caste, large landowners. When I was in Nilkanth, the village *panchayat* was unable to function. The chairman, a Kurmi elected in 1961, clashed with his antagonist, a Thakur, who had lodged a case against him in court, contesting his election. As will be seen later, this type of conflict is now taking very serious proportions in eastern UP and Bihar.

In the hamlets of the Chamars, most of them landless, the climate was particularly depressing. They earned one rupee a day, usually with some food. Permanent workers attached to a landowner received a plot of land for their own farming in exchange for their services. Thus, Dewan, a Chamar, had the use of 0.25 ha. of his Thakur landlord's land. He borrowed his bullocks to plough the land and remitted the straw in exchange. Dewan grew wheat in the *rabi*, followed by maize in the *kharif*. He was able to get a

[10] Bayly, *op. cit.*, p. 101.

yield of about 400-500 kg. of grain per year. His employer paid him Rs.4 per month, and one meal per day. His wife worked about four days a month for half a rupee per day and one meal. The whole family (three young children and an older son, married and working), barely survived, often with only one meal per day. They did, however own a cow which provided half a litre of milk per day for the children. They had also received a loan from their owner.

Casual labourers like Bulli, another Chamar, did not get enough work and survived with scant hope of ever escaping their lot. Like many others, Bulli was too weak, physically, mentally or intellectually, to react to and try and fight against his misery. It was almost as if, at a certain point, a very precarious and low standard of living blocks any road towards progress. 'Poverty does not leave the poor,' were Bulli's words.

When I left Nilkanth in March 1964, the situation was not desperate, but the immediate future looked rather grim.

The awakening

In 1975 and 1978, I revisited Nilkanth. Gradual but distinct changes in production were noticeable. Private tubewells began to appear and there was some growth in the consumption of chemical fertilisers and new high yielding varieties of cereals. New seeds of wheat were spreading faster than paddy, although here and there one would come across plots relying on new rice seeds. Potato and vegetable production was increasing moderately. On the other hand, sugarcane production remained as low as it was in the 1960s.

The process of growth accelerated between 1978 and our last visit in November 1985.

Yields of the main crops (kg./ha.)

	Paddy	*Wheat*	*Sugarcane (gur)*
1963-64	1,000-1,350	700-900	2,000-3,000
1978-79	1,300-2,000	1,400-1,500	3,000-3,500
1985-86	2,000-3,000	2,000-3,200	3,000-3,200

The consolidation of landholdings completed in 1978-79 had a clear impact on the accelerating construction of private tubewells. In 1985, there were 47 units, 21 belonging to Thakurs,

17 to Kurmis, 6 to Brahmins, 2 to Chamars, 1 to Yadavs (a middle cast more or less comparable to Kurmis). The two state tubewells each cover 40 ha. per season. Most private tubewells are 5 HP (investment cost Rs.20,000 - 25,000) with a command area of around 4 ha. each. Some of them rely on diesel because of the power shortage. Several owners also sell water to neighbours. The traditional wells (built along the *mot* system) have practically disappeared.

Around 190 ha. are irrigated by private tubewells and 80 ha. by state tubewells. Although irrigation is facing many difficulties (power cuts, for example), it has considerably improved since the mid-1960s, thus paving the way for the introduction of new seeds and chemical fertilisers. On the other hand, pest control, particularly necessary for paddy, is progressing very slowly compared to its use in the south-eastern deltas.

Some large farmers own tractors which they rent from time to time, but bullocks are still the predominant means of ploughing.

The progress of potato (6-24 t./ha.), and more recently of vegetables, chilli, and *haldi*, is particularly noticeable amongst the Kurmis, which means that the pull effect of the Varanasi market is now percolating down to the villages outside the ancient vegetable belt near the holy city. *Gur*, on the other hand remained stagnant.

Several farmers are also adopting a more intensive cropping pattern: wheat is followed by paddy.. A three month crop of potato follows immediately after the harvest of paddy. Sugarcane comes later. Others grow pulses in the *rabi*, maize in the monsoon, followed immediately by vegetables.

Animal husbandry has not made much progress and sales of milk outside Nilkanth remain limited.

The road to Nilkanth and further is badly metalled. The traffic has increased though not substantially: some *tangas*, bullock carts and clearly more bicycles, occasionally a tractor. It is worth mentioning that packed camels are still in use as in 1964. By midday a dozen are found loaded with fodder in the bazar of Mangari, and they return with bags of chemical fertilisers. In the narrow lanes, they coexist with buses, trucks, motor-cycles and cycle-rickshaws.

On the road approaching Nilkanth, tea stalls and tiny shops have been opened. Trade is expanding and some workshops have opened in Mangari. In 1978, Lohars (blacksmiths) in Nilkanth had started manufacturing iron ploughs, more efficient than the old plough, but light enough to be drawn by local bullocks. There was also a workshop of weavers working for a firm in Varanasi.

The advent of electricity has led to the opening of four flour mills. However, poor people like the Chamars still use the old mill stone at home. Small rice machines have been fitted to tubewells. Threshing machines for wheat, introduced in the 1970s, are quite widespread now. It was also interesting to note that by 1978 farmers had adopted the chaff-cutter moved by hand to cut the fodder, an operation formerly done with a small knife and hence more time consuming.

The construction of new homes remains much more limited than in Mirpur. One also sees fewer handpumps.

Though significant, the overall expansion of agriculture and of secondary and tertiary services remains at a level lower than in Mirpur, whereas the population growth has been faster: 67 per cent *vis à vis* 51 per cent in twenty years:

	Population of Nilkanth	*Density per sq.km.*
1961	2,500	600
1981	4,180	1,000

Regardless of the number of men working outside (for which I could not gather data), such a density has become dangerously high even when taking into account the latest growth of the economy. And this in a village relatively close to Varanasi. Although the rate of urbanisation was high in 1961, it has only risen from 22.5 per cent to 26.8 per cent (1981) for the whole district.

As for family planning, its effect remains mediocre. An attempt is now being made to speed up the propaganda. The motivator receives Rs.40. If he reaches a target of 15 cases of operated people he receives a television set (black and white, valued at Rs.3,500). Forty cases entitle him to a moped (Rs.5,500). Those who undergo the operation receive Rs.125 and lottery tickets and, more recently, a green card giving various privileges for obtaining a loan or for the education of their children.

Caste relations and standards of living

It seems that there are less instances of *dabao* (pressure) upon

Chamars victimised by landowners now as compared to the 1960s and 1970s. One comes across more Scheduled Caste people who assert themselves. Some of them are now a little better off, more educated and more conscious of their rights. Yet, it is still unacceptable for a Chamar to sit on a *charpai* (common Indian beadstead) next to a twice born, as has been permissable for some years now, in Mirpur. Some poor families have been given a tiny plot of communal land for their homesteads.

The Kurmis continue to strengthen their position through hard work. Rivalries with the Thakurs are far from over and now the latter, like other high castes, are catching up economically. In 1978 itself I could see how they were breaking the old taboo of not touching the plough and were becoming more involved in work on the fields. As far as the rather big landowners who have their land cultivated by sharecroppers (*batai*) are concerned, they bear half the cost of chemical fertilisers and pesticides (rarely used) but water and seeds are paid for entirely by the tiller.

Until 1978, daily wages in real terms remained more or less stagnant, whereas a clear increase was noticeable in the early 1980s:

Daily agricultural wages

	In cash (Rs.)		*In kind (wheat or paddy [kg.])*
1964	1.0	with some food	2–3
1978	2.50–3.0	often with some food	3–4 rarely up to 8
1985	10.0	often with some food	4–6 up to 8[11]

While a number of landless labourers do benefit from rising wages and, not infrequently, from increased employment facilities, there are still people whose lot remains miserable. In certain cases, they have to work for lower wages because of pressure from the landowners, or, like the Musahars (Scheduled and semi-migrant caste) they remain so poor that they can hardly involve themselves in the process of growth. As in 1978, I spent some time around their shabby huts built with leaves and bamboo. The small children walk around naked with their ballooned stomachs, a clear indication of malnutrition. The huts are almost bare. The Musahars

[11] Retail price of wheat per kg. in November 1978 was Rs.1.30-1.40. Rice stood at Rs.2. In November 1985, wheat was Rs.2 and rice, Rs.2.50-4.0.

are underpaid. Being few in number (about twenty), they do not have the relative bargaining power of the larger Chamar community. On the other hand, the Thakurs believe that the Musahars are very 'independent minded. They do not want to work and smoke *charas* instead!' (a drug similar to hashish).

Bewati, a Chamar, has no land except the few square metres of his homestead. Two of his three sons go to school and two of his four daughters are married. He complains that he does not always get the regular wages. He may be compelled to work for only 3 kg. of wheat or paddy per day. His house is in a poor condition and bare except for a few beds. He has only one set of clothes. I do not know how he managed to do so but he has bought a cycle-rickshaw. He earns Rs.5 for every three persons transported to Mangari. He also has one bullock and a small cow.

Other Chamars join us. Some of them admit that their life is improving at last. Many more children have been educated. There is some additional work in the village. Educated Harijans can obtain a job under the reserved quota in administration. Another Chamar has only 0.2 ha. Until recently he used to grow wheat and maize. Now he is able to grow potato in between. He would like to replace maize with paddy but for the insufficient supply of water.

Ram Das is a Shudra. He lives with his wife and their two adult — unmarried — sons and two young girls. Six children died in their early years. At my question: 'have you got land' he answered, 'not even for the call of nature,' but he does own his own homestead. As Ram Das suffers from asthma he is too weak to get more than 3 kg. of wheat as a share of the harvest unlike stronger men who get double the quantity. The two sons work as casual labourers but they seem to be doing quite well as each of them owns a cycle

Jhulu, belonging to the lower caste, Rajbhar, is more fortunate even though he owns no land. His eldest son works in Bombay. Another is married in the village and four younger children live at home. With his pair of bullocks he earns Rs.25 per day for ploughing the fields. Jhulu and his wife also work as labourers, transplanting and harvesting paddy. They are busy during the wheat harvest. Jhulu is expecting a second set of clothes from his son in Bombay, and his wife a third set. These clothes come in addition to Rs.500 every year. True they get more work, but previously, when they had no children, concludes Jhulu, life was easier.

Let us move now to the Kurmi hamlets. Lalchand Prasad lives with his parents, his brother and his family and the families of his two other brothers who work in textile mills in Bombay. In all, around twenty-five people — adults and children — live off 1 ha. of land. On 0.25 ha., paddy (800 kg., equivalent to 3,200 per ha.) is followed by 600-700 kg. of wheat (2,400 - 2,800 per ha.). They have a small plot of sugarcane giving 400 kg. of *gur* which is sold for Rs.1,600. On other parts of their land they grow maize, pulses, potato and vegetables. The latter two are sold for Rs.1,000. The family has a pair of bullocks with which they occasionally plough fields for others. Lalchand and his brother also work as agricultural labourers earning Rs.10 per day. They also earn some money from petty trade, buying and selling vegetables. In addition, the family receives Rs.2,500 per year sent by the two brothers in Bombay. Thus, they manage quite well in view of their small holding. They have one bicycle, a radio and a few sets of clothes per person.

Their neighbours are also a large family of twenty-five people, living off 1.50 ha. with a similar cropping pattern. They own a tubewell jointly with another family. However, they have few external resources. Only one young man is working outside as a compounder at the block headquarters.

Without their solid traditions of hard work and vegetable gardening, and without the supply of new inputs including better irrigation, such people would be able to survive only with great hardship.

The high castes who had remained for so long rather unenterprising, have begun to show several signs of change. The medium and upper Thakurs and Brahmins have invested in tubewells, a great help for the *rabi* crops, as well as for paddy which, on large tracts, is no longer rainfed but irrigated. No less striking is their — at least the small farmers — more active involvement in activities like ploughing.

Lalji is a Brahmin, now fairly old. With his two brothers they own 1.25 ha. of land, growing cane, paddy, wheat and potato. They have a pair of bullocks and three cows which together supply 1.5 - 2 l. per day. Life is hard no doubt, but the rising yields of crops enable the whole family to lead at least a frugal existence.

His neighbour, Vijay Shankar Trivedi lives with his aged

parents, his children and some members of his brother's family. The household consists of fifteen adults and children on a holding of 2.4 ha. irrigated by their own tubewell. On the whole they are rather well off because three brothers are employed outside: two as teachers, one in the army, and all three send money home.

We met one of the sons of Kamla Prasad Singh, the main Thakur landowner who, even in 1964, was modernising his holding. His crops have increased considerably due to tubewells, new seeds and chemical fertilisers. Most of the land is cultivated with the help of sharecroppers. They meet half the expenditure on chemical fertilisers and pest control and the entire cost of irrigation. Yields of paddy are not particularly high: 2,400 kg./ha. Some of the land suffers from excess water. Wheat does better at 3,000 kg./ha. and the yield of potato is 24 t./ha. An orchard of mango and gwava trees is rented to an outside merchant.

These changes are no doubt encouraging, but one must consider, on the other hand, the growing sub-division of holdings. Already there are households of 15-30 persons who rely on 1 ha. or so, which will be further divided when the head of the family dies. No doubt the agricultural potential has not been fully tapped. There is room for pushing yields further upwards and to widen the process of crop diversification, particularly in favour of vegetables, chilli and spices. There is a large market in Varanasi and already consignments of vegetables are sent to Calcutta.

To achieve these goals, we are once again confronted with the problem of water management. More tubewells are needed provided the supply of electricity improves, a prospect rather doubtful in the immediate future. Otherwise farmers will have to rely on more diesel engines, a less profitable solution. Proper drainage and flood control, a problem less acute here than in Bihar, are also needed.

A number of families are able to make both ends meet when some members work outside agriculture and/or outside Nilkanth. It is crucial to enhance this process, but how far is it possible? The data gathered at the district level (see later) are not very encouraging.

Coming to family planning, over the last twenty years an awareness of the issue has percolated down as in so many other villages, but the programme proceeds only gradually.

Finally, how are the schemes geared towards the poor operating?

Although I could not gather as much evidence as in Mirpur where both positive and negative aspects are noticeable, I came across many complaints at all levels: from the poorest to the richest people. One Chamar complained that for an IRDP loan-cum-subsidy, he had to give a *baksheesh* (*ghus* as they call it) to the bank clerk. Another Chamar says that all the money goes to the rich. The poor and illiterate like him can do nothing. '*Bhare log khate hein*' (the rich people eat), the poor have nothing, adds a third. A small landholder (with 0.8 ha.) is entitled to a loan-cum-subsidy, but the cooperative society delays matters until some '*ghus*' is given. A Kurmi complained that to obtain a clerical job in another block, he had to pay Rs.3,000 to the BDO and the ADO and adds: 'If I was a Thakur, I would have to pay less.'

Varanasi district

The population has increased from 2.35 million[12] in 1961 to 3.70 million in 1981, i.e., 57.4 per cent. The geographical area is 5,091 sq.km. and the density 725 per sq. km. We have to take into account the high rate of urbanisation : 22.5 per cent 1961, 26.8 per cent in 1981. In 1985, Varanasi town had a population of around one million. In the neighbouring less urbanised areas, the average district density falls to 600-630 per sq.km.

In the past twenty years the net cultivated area has remained practically unchanged but the gross area (including multi-cropping in the same year) has increased, though not much in the past seven years:

	1984-85
Net cultivated area	323,000 ha.
Area cropped more than once	146,000 ha.
Gross area	469,000 ha.

	1976-77	*1984-85*
Irrigated area	175,900 ha.	224,000 ha.
Gross Irrigated area	221,000 ha.	289,000 ha.

In 1984-85, out of the total net irrigated area, 117,000 ha. depended on tubewells and 96,000 ha. on canals.

[12] Data supplied by the district administration. See also *Varanasi Suchna evam Jan Sampark Vibhag*, 1985 (Hindi).

Paddy is no longer the main crop in the district (143,000 to 155,000 ha.) when compared with wheat which increased from 146,000 ha. in 1980-81 to 174,000 in 1984-85. In the last few years oilseeds have expanded from 47,400 ha. to 88,000 and pulses from 17,400 to 76,000 ha. Sugarcane amounts to only 13,400 ha.

The consumption of chemical fertilisers has nearly doubled between 1977-78 and 1984-85, reaching 63,750 t. (in terms of nutrients).

Average District Yields (kg./ha.)

	1963-64	*1975-76*	*1982-83*	*1983-84*	*1984-85*
Wheat	700-900	1,280	1,440*	1,720	1,550
Clean rice	600	880	1,370	1,420	1,330
Paddy	900	1,350	2,000	2,100	2,000

*there were practically no winter rains.

The average data are substantially lower than that recorded in Nilkanth, a difference which was already noticeable in the mid-1960s. The progress of irrigation in Nilkanth is probably higher than in other parts of the district where non-irrigated wheat yields around 1,000 kg./ha. and rice, 800-900 kg./ha. (1,200-1,350 in terms of paddy).

All these data confirm what is observed in other parts of eastern UP. The long period of a semi-stagnant economy is over as a result of more dynamic trends: in particular, tubewells, rising consumption of chemical fertilisers and new seeds. The impact of growth on wages as observed in Nilkanth is no less significant.

I was unable to get data on the break up of agricultural holdings for the district. The data for Varanasi and the surrounding districts confirm the trend of growing division of holdings and the rising number of landless labourers. Out of 4.25 million operational holdings in the districts of the eastern plains of UP (including Varanasi), 81.6 per cent were below 1 ha. in 1980-81, another 11.3 per cent between 1 and 2 ha. Both categories covered 60 per cent of the cultivated area. About 55 per cent of the active population of these districts are cultivators and 20 per cent agricultural labourers.[13]

[13] Reserve Bank of India, Report of the Committee on *Agricultural Productivity in Eastern India*, 1984, vol. II, p. 368 and 376.

Total Working Population in Eastern UP (million)

	Total Workers	*Cultivators*	*Agricultural Labourers*
1971	10.4	5.7	2.9
1981	11.9	7.1	2.4

Source: Reserve Bank of India, *Report of the Committee on Agricultural Productivity in Eastern India,* 1984, Vol. II, p. 415.

The fall in the total number of agricultural workers is surprising. Does it imply considerable out-migration and thus many more jobs outside agriculture?

Regardless of the accuracy of these data, the process of growth needs to be considerably strengthened in view of the existing poverty and the future division of holdings. At present nearly half the gross cultivated area is under irrigation. Of the total ground water potential, 65 per cent has already been tapped, which does not leave much room for further expansion. However, sizable gains could be achieved with better irrigation, which brings us back to the problem of power. As far as surface water is concerned (canal irrigation), I do not know what more can be done to expand it. On the other hand, the classical shortcomings found in the south are not lacking here either: there is a need to replace field to field irrigation in canal areas by more field channels and outlets; to encourage a more careful use of water, etc.

Many parts of the district are exposed to excess water and floods as observed in Nilkanth. A variety of paddy (*mahsuri,* 110 cm.high) can stand a relatively high amount of water; its maximum yield will level off at around 2,500-3,000 kg./ha. of paddy, already a very good result compared to the yields of old varieties of approximately 1,000 kg./ha. Yet, in well drained, well irrigated fields, yields of 4,500-5,000 would be possible.

Progress in milk remains limited, especially when compared with the north-west, but even in the 1960s, Varanasi and other eastern UP districts were well behind the more progressive areas.

Afforestation schemes through private nurseries as well as the forest department have gathered momentum to protect orchards and fuel wood. In 1984-85, 4.2 million trees were planted in the district with a mortality rate of at least 30 per cent.

The various avenues leading to higher yields, the shift to richer crops like vegetables, the expansion of non-agricultural activities, especially around big villages and small towns etc., would do well to proceed faster and thus create more employment, because one cannot rely too much on the absorption capacities of Varanasi. The city absorbs some of the surplus population of the district, but it also attracts people from many other districts. Besides, the growth of urbanisation has not been spectacular in the past twenty years, although the base year itself already had a high percentage of town dwellers.[14]

When discussing all these issues with the district authorities, I gathered the same impression as I did in Bulandshahr. Development administration is gaining roots. One now meets more senior officials who know their facts and data, than one did in the 1960s or even in the 1970s. Planning is more organised and the offices of the state services do have a clearer idea of the major requirements and constraints.

The District Rural Development Agency (DRDA) is in charge of the various schemes designed to help the poor. In spite of some of the drawbacks referred to by the people of Nilkanth, one cannot write off the whole scheme of loan-cum-subsidy, but the costs are high. In 1984-85, the IRDP covered 19,234 families in the 22 blocks of the district at a cost of Rs. 62.3 million, including 29.7 million as subsidy, the rest as loans. The loan-cum-subsidy programme is available for farmers with holdings below 1.2 ha. Institutional reforms (consolidation of holdings), technical innovations, changes in attitudes and behaviour have finally brought to an end the long period of decay which began in the 19th century. This is a glaring example of where one needs a rather long term perspective in order to avoid shaky judgements. By 1964, in Mirpur one did expect further progress fairly quickly, though not to the extent to which it actually occurred. In Varanasi, on the other hand, one had the feeling of an almost total deadlock, but this is no longer the case.

[14] This rate of urbanisation in Varanasi is not typical for the region. In 1981, 89.4 per cent of the whole population of eastern UP lived in rural areas (92.6 per cent in 1961) and 80 per cent of the total workers were engaged in agriculture (Reserve Bank of India, *op. cit.*, p. 340).

Chapter Eight

Acute Poverty in a Fertile Land

For at least twenty years, the image of Bihar has been dark: poverty, bitterness, political instability, increasingly bloody feuds, weak administration, growing population pressure, growing deficit in foodgrain, etc. What has gone wrong in these fertile plains which, for centuries, had been the cradle of Indian civilisation? Patliputra (near Patna) had been Ashoka's capital in the 3rd century BC. Later, at various periods the area had been quite prosperous. This was still the case with the districts south of the Ganges in the 18th century and after. Although the plains north of the river suffered from relative isolation, they too were not, until the end of the last century, considered the poorest areas in India. According to the Gazetteer of Darbhanga (a district adjacent to Muzaffarpur):

> The innumerable rivers and rivulets, the alluvial soil and the usual monsoon rain make the tract one of the most important production centres of Bihar and it has rightly been described as the granary of India. The climate is responsible for the characteristic ease loving complacent attitude of the people who are allergic to hard work.[1]

It is probably true that with a low population, fertile soils and a high rainfall, life must have been relatively easy. There was no need for irrigation and vast tracts of land could bear a second crop in the dry season with the moisture left in the soil. The level of population being low, it was also possible to leave the areas exposed to floods uncultivated, a common phenomenon, especially in north Bihar. From time to time, however, a major failure in the monsoon could lead to a severe famine as in 1874.

[1] P.C. Roy Chaudhury, *Darbhanga* (reprint), Bihar District Gazetteer, (Patna, 1964), p. 24.

The land-man ratio and the population-resources ratio began to be gradually eroded. In 1876, the density per sq.km. touched 289 in Muzaffarpur district with a population of 2.2 million. After a period of stagnation, as found in many parts of India between the end of the 19th century and 1921, the pressure of the population increased. As can be expected, all available land has been reclaimed, including land exposed to floods or at least excess water, which increased crop variations from year to year. Irrigation hardly expanded at all. No new varieties of seeds, no new techniques appeared before Independence.

On the other hand, indigo planters and European planters had played an important role in Muzaffarpur from the end of the 18th century. European planters encouraged the building of roads, horse breeding and had some fine cattle farms. The decline of indigo production between the two world wars impoverished the economy. It was only partly replaced by sugarcane, that too with a low yield. As for the urban economy, it grew only very gradually. The urban population was below 5 per cent in the district.

The land tenure system represents another important factor that can explain the present poverty. Bihar depended on the Permanent Settlement of Lord Cornwallis (1793). Large estates belonging mostly to Bhumihars and Thakurs were created. Although many of them were later divided and tenants received occupancy rights during the last century, the Gazetteer of Muzaffarpur notes (around 1900), that landlords were 'unusually powerful' and the tenants 'unusually depressed'. The labourers were in an even worse condition receving 2 kg.of low quality grain per day.[2]

Kalpi village

The village is located along a good district road, 20 km.east of Muzaffarpur. The landscape is completely flat with paddy fields and woods surrounding the hamlets (distributed castewise). The old (*bhuri*) Gandak and other rivers flow around the village.

The cropping pattern found in 1967 hardly changed for many decades. Paddy was — and remains — the dominant crop. On

[2] P.C. Roy Chaudhury, *Muzaffarpur* (reprint), Bihar District Gazetteer, (Patna, 1958), pp. 190, 247.

higher lands there was some maize and unlike in Nilkanth, there were no private wells for irrigation. Some years ago, a state tubewell had been constructed. In spite of poor maintenance and functioning, it irrigated some plots of land where, with the help of block officials, a few farmers had started growing new varieties of paddy. Otherwise, paddy was rainfed giving a very low yield (1,000-1,500 kg./ha.). Several tracts of low land suffered from excess water due to the lack of proper drainage, so that yields here were even poorer, often below 1,000 kg./ha. In the *rabi,* some land would bear unirrigated wheat and pulses. One came across few cash crops: small plots of chilli and tobacco.

The dominant impression was of a stagnant economy with a growing population. Besides, the social climate was depressing: misery and bitterness on the part of Scheduled Castes, landless labourers and small farmers, and, the lack of an enterprising spirit among the few big landlords, all led to tensions, and abuses committed by the upper farmers.

However, one could notice a change of attitude. Bihar had just been hit by two consecutive devastating droughts in 1965 and 1966. But for the abundant supply of American wheat and its efficient distribution by the Bihar administration[3] and several voluntary agencies, Bihar would have known one of the worst famines in its history. Although there were very few casualties the drought made the people conscious that the old rainfed agriculture had to change. Whether it was in government offices or in villages, I got the same answer to my question about the most urgent priority : 'water, irrigation.' This was needed as a protective measure to compensate for the vagaries of the monsoon and to push crops in the dry season which would give a higher yield than if they were to rely only on the moisture left by the monsoon and on a few scattered winter rains. A number of upper and medium farmers had applied for a loan to construct a tubewell. The Bihar government and the district administration were making plans to encourage irrigation where it was non-existent and to speed up the major irrigation works under construction like the Kosi and Gandak projects.

[3] The central government sent some of its best senior civil servants who, in spite of the inefficient local political leadership, organised relief works and supply of grain to the villages: free rations to the poorest, sales at a fair price to others.

In 1978 and 1985 I revisited Kalpi and the district. No doubt agriculture is no longer stagnant, but growth falls much behind Nilkanth.

Kalpi

Geographical area		479 ha.		
Net cultivated area (1967 – 1975)		395-400 ha.		
(1985)		370*		
Irrigated area (1985)	...	*Kharif*	...	75 ha
		Rabi	...	135
Population	...	(1961)	...	2,471
		(1981)	...	3,608
Density per sq.km.	...	525	...	762

*Some land has been used to construct an embankment for flood protection on the old Gandak.

It is particularly difficult to estimate average yields in view of the great variations of crop conditions. The low-lying lands remain, as before, submersed by so much water that paddy is sown broadcast. Thus, on 50 ha. the yield of paddy may be as low as 500 kg./ha. On slightly higher lands, 100 ha. remain rainfed with a stagnant or fluctuating yield. If there is neither drought nor excess rain the harvest will be around 1,500 kg./ha., otherwise it falls to 1,000. Approximately 50 ha. are irrigated on relatively high lands, yielding around 2,500 kg./ha.

Wheat cultivation has clearly increased with the introduction of tubewells. Out of the 240 ha.devoted to wheat, half is irrigated providing 2,500 kg./ha., sometimes only 1,500, sometimes as much as 3,000. Non-irrigated wheat yields 500 to 1,000 kg./ha. The yield of maize varies according to the rains: 1,000 to 1,500 kg./ha. The yield is higher when grown in the dry season.

In the low lands, it does not make sense to use chemical fertilisers because they can be washed away by too much water. New dwarf varieties of paddy are of course out of the question. In the intermediate areas, chemical fertilisers could pay, but there is some risk involved and that is why farmers hardly make use of them. The same is true for non-irrigated wheat.

Irrigated crops do benefit from chemical fertilisers, but judging from the yields, the applications are substantially below the recommended doses. As for pesticides, they are hardly noticeable though they are badly needed for the monsoon crops.

Vegetables, chilli and tobacco are on the increase. Some rich Kalwar farmers (see next section) as well as small farmers have either a large field or small kitchen gardens. On 0.4 ha., chilli and vegetables bring a gross return of Rs. 10,000 and a net return of Rs. 5,000 per year; 0.4 ha. of tobacco bring Rs. 4,000 (net).

Animal husbandry, especially milk, continues to play only a minor role in the economy. A cyclist carrying milk on the main road is not a common sight.

Seven landlords have each bought a tractor, which they rent occasionally. Yet, a sizable amount of ploughing still relies on bullocks.

Water management remains the most fundamental constraint, and a very complex one : large scale drainage works are needed; flood protection is also necessary and the embankments on the old Gandak have not been completed. Irrigation should be made available to the whole area. There is plenty of ground water (20-30 metres deep). Yet, in 1985 the state tubewell was operating as badly as before. In 1978 it had broken down altogether for some months, because some people, during a power break down, had cut and taken away 30 metres of electric wire! In addition there are thirty-five borings of tubewells using twenty-two pumpsets which can be shifted from one boring to the other. The supply of electricity is so unreliable that all the pumpsets work on diesel.

A few flour and rice mills have been opened in the past twenty years. Some more shops have appeared along the road since our first visit. Some new houses have been constructed, one of them is a spacious mansion in concrete which belongs to a rich Kalwar (Banya). The houses and courtyards of the Bhumihars have not changed, nor the huts of the Chamars and Dusads.

Trade has expanded, but, despite the good road, the traffic to and from Muzaffarpur remains moderate compared to more progressive districts : there are only a few trucks and buses and bullock carts.

Out-migration does play a role in the village economy and helps a number of families as will be seen later.[4] Unfortunately, I could not gather even rough estimates. Data for the district and the state do not suggest a very strong process of diversification of the

[4] In addition to the age-old migrations to Assam and Bengal, some people are now going to Punjab.

economy which would widen the employment market in the secondary and tertiary sectors. Hence, the number of small landowners and landless labourers keeps on increasing.

Incomes and social relations

On the basis of such a modest economic growth it would be futile to expect much improvement in the standard of living of the poor. Let us first look at the trends in wages.

Daily wages of agricultural labour

	In cash (Rs.)	*In kind (kg.)*
1967	1.50 and some food	not available
1978	2.50 – 3.00 and some food	2.50 – 3.00-4.00
1985	5.00 – 6.00 and some food	3.00 – 6.00

Daily wages can be paid in cash or in kind, or both. The cash wages at Rs.3 in 1978 showed no decline in real terms with reference to 1967. Between 1978 and 1985, wages have more or less remained constant in real terms, or have shown a very slight improvement. Wages in kind vary. For land tillage and transplanting (1985), daily wages can amount to 2 or 2.50 kg. of grain (wheat, paddy or some inferior grain), some employers providing one meal others not. But there may be cases where the wages may fall to 1.50 kg. of grain and one meal. At the time of harvest wages are paid in kind. The worker keeps a share of the harvest; for paddy it varies around one-tenth or one-ninth of what the worker harvests daily. Of course, the lower the yield, the lower the wages. Out of well irrigated plots, one man can earn up to 10 kg. per day, but it is rare: the most common range being, as mentioned above, 3 to 6 kg.

Many Bhumihars and Brahmins keep attached labourers. The arrangement is that the worker is allotted a tiny plot of land for his own use or on a sharecropping basis. It can be 0.04 to 0.2 ha. Such labourers receive, in addition, daily wages, usually in kind (often inferior grain), amounting to 1.5 – 2 kg. The labourers 'enjoy' a minimum of security compared to casual workers, but their living conditions remain very precarious.

While we have seen the active role played by women in agriculture in south India and their importance as wage earners, in Kalpi as in most villages in this area, even for very poor and low caste women, 'it is not done' to work for others. One may find women working in their own fields, but not elsewhere, a custom which makes the lot of the poor even worse and can also contribute to low yields. In 1967, for instance, the monsoon was delayed and paddy should have been transplanted in less than the normal time. But this could not be done with only male labour!

Social conditions as seen from an economic angle are already bad enough. To the hardships caused by the imbalance between economic growth and population, one must add the semi-feudal relations prevailing between the rich and the poor. Unlike in many other parts of India, one finds in Bihar, and particularly in the northern districts, very large landlords still owning 50-100 or even 200 ha. Some of them can be found in Kalpi. No wonder then that the word *dabao* (pressure) is often heard in the Harijan hamlets. Bitter complaints are raised against 'the fat' *(mota)* 'who sit idle while the Chamars work'. I remember when in 1978, I was interviewing some Chamars, two young Bhumihars, sons of landlords, dressed in fine, spotlessly clean *dhotis*, were passing by on their cycles. They stopped to make fun of me, a foreigner wasting his time with these Untouchables!

Conflicts are, however, not confined to high caste landlords versus landless Harijans We find here again our good and enterprising Kurmi and Yadav cultivators who enter increasingly into conflicts with the high castes. At the same time they may also bully the very poor. In Kalpi and in other villages, I did not hear of bloody and armed clashes between these three sets of castes/classes as one does in south Bihar, but the social climate is tense.

Let us now look at some individual cases. Chalita Ram, a Chamar, lives with his wife, his five young sons and one daughter. They don't own any land. Even their homestead belongs to their landlord. He has been allotted 0.06 ha. of land for his own use which brings him about 60 kg. of paddy followed by 80 kg. of wheat. One of the sons works as a casual labourer for Rs.5-6 per day plus one meal of maize. Chalita Ram is quite alert in spite of his acute poverty. He has no other clothes than the ones he wears. Near his house stands his half broken cycle. Life goes on even though Chalita Ram is barely able to survive.

Binder Ram, another Chamar, and Ram Pasman, a Dusad (Scheduled Caste), work under a similar arrangement for their landlords. The latter has six surviving children. When he is lucky, Ram Pasman digs out 10 kg. of paddy stored by the rats in the field bunds. He also catches water crabs, some of which are eaten. The rest are sold at Rs.6 per kg. One of the sons works as a permanent labourer for another landlord.

Moti Ram, a Chamar, has five surviving children, three died very young. He owns his homestead and does casual work, like his wife. One of his daughters was married at a cost of Rs.2,000 borrowed from various sources. Moti Ram has benefited from the IRDP, getting a loan-cum-subsidy (total Rs.2,000) to buy a cycle-rickshaw. Of this Rs.400 went as *baksheesh* to the bank officials and despite this the bank supplied a cheaper rickhshaw than stated in the loan-cum-subsidy application! In addition to agricultural work, the rickshaw could have been a great help. Unfortunately, in the rainy season it is difficult to find customers and the road is often flooded. However, Moti Ram is a little less poor than the others: he has two sets of clothes.

Amongst the small farmers, one does not only meet lower caste landowners. Lakshman Mishra's father, a Brahmin, owned 5 ha. which were lost either in litigations or sold in order to get his daughters married. Now only 0.15 ha. are left, supplying a few quintals of maize and wheat. And yet, Lakshman lives reasonably well. His four sons have found jobs outside and have taken their families with them, sending money home to their father.

Ram Chandra Jha, another Brahmin, has only 0.4 ha. of unirrigated land yielding 400-600 kg. of paddy (1,000-1,500 per ha.). His fields suffer from too much water to be able to yield more in the monsoon. The moisture left in the soils enables a wheat crop of 400 kg. (1,000 per ha.). Like his neighbour, Ram Chandra nevertheless manages to live, but rather frugally, and his house is shabby. One of his sons works as a clerk in Delhi, another at the railway station of Samastipur in the neighbouring district. Both their families live in the village.

A Thakur in the other hamlet owns only 0.3 ha. Since the land is near the river, it is frequently flooded. Four of his children have died and only two survive. Though aging, he works from time to time as a labourer.

Krishna Chandra Jha is better off with 4 ha., half of which are irrigated. He has made a boring and rents a pump, harvesting 2,200 kg./ha. of paddy, as against 1,000-1,500 on his rainfed fields. Irrigated wheat on 1.2 ha. gives 2,200 kg./ha. and 0.4 ha. of hybrid maize (dry season) provides him with the equivalent of

4,000 kg./ha. His two bullocks are used for ploughing. The two cows each give 4 l. per day, an unusually high yield. The two older sons have got good jobs in Bangalore and Bombay and each sends home Rs.400-500 per month to maintain their families in Kalpi.

The Kalwars (Banyas) are on the whole more enterprising than the Bhumihars, Brahmins and Thakurs. Ram Chaudhury, an only son lives with his old parents, his wife and their three children. They own 2.6 ha., part of it irrigated by their own tubewell. In addition to paddy and wheat (each yielding 2,500 kg./ha. when irrigated), he devotes 0.6 ha., to vegetables. This year (1985) he has been unlucky because the crops were damaged by heavy rains. But under normal conditions he can earn Rs.8,000 (net). Ram Chaudhury has bought a second-hand tractor for Rs.27,000 which he uses for his own fields and rents out for ploughing or transportation.

Raghunat Chaudhury, of the same caste, does equally well. Although the father has divided his 12 ha. of land between his five sons, all of whom live in the village, they all work together and jointly own one boring and rent a pump. They cultivate wheat, paddy and vegetables with the help of casual labourers. They have now gone into business, buying and selling grain and chemical fertilisers.

In 1967, it was particularly distressing to visit the main Bhumihar landlords, one with 40 ha., the other with 50 ha. in Kalpi and 80 ha. in other villages. They were reluctant to invest in their land in the belief that with such large holdings, they could live comfortably despite very low productivity. In 1978, this situation had hardly changed, whereas in 1985 they had finally begun to be more active. One of them had by then dug three borings upon which he shifted his pump so that part of the fields could be irrigated. However, yields could have been, with the full package of inputs, higher than the 1,800-2,000 kg./ha. of paddy he obtained. He also had a thresher and a tractor.

The breakthrough is still to come

Economic stagnation in Kalpi has come to an end, but progress remains too slow to percolate down to the poor, except for a few of them. Is their standard of living declining? Perhaps not, but it seems doubtful that it is increasing. The social tensions, the semi-feudal relations between landlords and labourers have not changed much since 1967. However, the Bhumihars have, to a certain extent, lost part of their influence in the face of the more enterprising Kalwars who have bought some of their lands.[5].

In spite of some progress, the major constraint by far remains poor water management. Low lands should first be drained so that they can be irrigated like the rest of the area. New seeds, chemical fertilisers and pesticides should have been introduced on a much larger scale by now. One can easily imagine the enormous economic and social progress which would result. Even holdings of half a hectare would give two or even three good crops a year enabling their owners to lead a much better life. As for tiny farmers and landless labourers, there would be more employment for them and wages would rise in real terms. One could also expect chain reactions in the secondary and tertiary sectors leading to more employment.

The special schemes geared towards the poor provoke similar criticism, both from the poor and from the rich, as recorded in Nilkanth. Although some benefits do reach the poor, there are numerous stories of corruption similar to that of our Chamar friend who wanted to buy a rickshaw.

Unfortunately, I could not gather data on the labour force finding jobs outside Kalpi. This is rather important even though the poorest people find it more difficult to benefit from it. Is it a matter of education or of caste relations? Probably both factors play some role.

The view from the block and the district

How do the socio-economic conditions at Kalpi fit into Muraul block and Muzaffarpur district?

The cultivated area of the block (13,000-14,000 ha.,net) has remained practically unchanged over the last twenty years, but the irrigated area which was negligible in 1967 reached 3,040 ha. for the *kharif* and 7,750 for the *rabi* in 1985. There are now twenty state tubewells (three of which are out of order), 1,880 borings for private tubewells depending on 120 electric pumps and 910 diesel

[5] Observations made by Gerry and Jeanine Rogers during a visit to Kalpi in 1982.

pumps. The progress of irrigation mostly for wheat in the *rabi* is worth emphasising. Wheat covers today 6,000 ha., as against 7,300 for paddy, 2,130 for maize and 2,130 ha. which are devoted to other crops.

Yields for the main crops for 1984-85 (Muraul Block)

	(kg./ha.)
Paddy (local variety rainfed)	1,100 – 1,300
Wheat	1,600 – 1,900
Maize (*kharif*)	1,100
(*rabi*)	3,200
Potato	12,000

	1961	*1981*
Population	83,587	116,545 (+ 40 per cent)
Density (per sq. km.)	483	673

At the district level, comparisons of population and production are made difficult by the bifurcation of the district between 1971 and 1981. The present net cultivated area is around 240,000 ha. (gross 354,000 ha.); 62,000 ha. are irrigated in the *kharif* and 85,000 in the *rabi,* a substantial increase when keeping in mind that the district (before bifurcation), had only 8,400 irrigated ha. in 1966.

	1966-67	*1977-78*	*1984-85*
Number of state tubewells	96	194	252
Number of private tubewells	201	12,515	22,991

Most of the state tubewells are electric, but their performance is poor due to the shortage of power and poor maintenance. The private tubewells depend on 3,258 electric pumps and 10,965 diesel pumps. They also face various difficulties in terms of maintenance and availability of parts. Some of the pumps made locally are not of a good quality.

Wheat has been the main beneficiary of irrigation (around 20,000 ha. in 1966-67) now covering 75,000 ha., but irrigation and the supply of other inputs suffer from shortcomings since the average yields amount to 1,400-1,500 kg./ha. Paddy (which accounted for 150,000 ha.) fell to 900 kg./ha. in the drought year of 1983-84 as against 1,300 in a better year like 1982-83.

These low yields can be explained by the relatively large amount of land which remains unirrigated, the fact that irrigation is not adequate, and by the rather low consumption of chemical fertilisers which amounted to 14,800 t. in 1984-85 (in terms of nutrients)[6] versus 13,300 t. in 1977-78.

The global process of rural development is far from impressive. Agriculture is expanding too slowly compared to the population. In 1981 the total population of the district (after bifurcation) had reached 2.35 million with a density of 747 per sq.km. The rate of urbanisation has only progressed from 4.6 per cent in 1961 to 8 per cent in 1981. Towns are expanding slowly as are trade and small and medium industries. Marketing and transport facilities are poor or at least inadequate. Workers have to go far away to Bombay, Calcutta or Punjab to find jobs. As a result, as many as 80 per cent of the district's active population is involved in agriculture: 265,000 cultivators and 273,000 agricultural labourers, a situation unchanged over the past two decades.

The moderate performance of agriculture prevents a decisive upsurge in the other sectors of the economy. But why is the performance mediocre? In fact, in an official note on the district dated 1966, one finds that: 'The district on the whole has very favourable soil and climatic conditions for the growing of crops and fruit'.

From Kalpi to the district level, the population pressure has become very heavy and the number of agricultural labourers particularly high. These conditions require a much sharper increase in production than has been witnessed so far. Although some large estates remain, the division of holdings keeps on increasing. For the whole of north Bihar, in 1981, 85 per cent of the holdings were below 1 ha., i.e., 350,000 versus 190,000 in 1971.[7]

Water management, again, is the major constraint. Irrigation is badly needed and is possible on a much larger scale than it is now. But massive drainage and flood control works are also required. In fact, the plains of north Bihar receive an enormous amount of water from the rivers coming down from Nepal which seem to concentrate on a relatively small area so that few regions of India

[6] Reserve Bank of India, *Report of the Committee on Agricultural Production in Eastern India*, 1984, vol. II, p. 263.

[7] Reserve Bank of India, *op. cit.*, p. 263.

face problems of that magnitude. In Muzaffarpur and in the other districts of north Bihar, nearly one-third of the cultivated area can be flooded during the monsoon, and, as a result, must be left uncultivated.[8] Water management is therefore much more complex than in Punjab and western UP where it was enough to have enterprising farmers sinking their tubewells. That is more difficult in Bihar, though by no means impossible, for other reasons as well. The number of very small holdings is extremely large and there has been no consolidation of holdings, an obvious incentive in UP, Punjab and Haryana. Drainage and flood control works are beyond the capability of private farmers. They depend on large government projects. Bihar is not lacking in major irrigation and flood control projects which have been in existence since Independence, but the results have so far been too slow, if not disappointing.

The use of chemical fertilisers, extension services, the supply of seeds and, credit facilities have been rather inadequate over the past twenty years, a situation that remains unaltered even now.[9]

Land records are not up to date and are badly kept. As I was told in Muzaffarpur, it seems extremely difficult to consolidate land-holdings. Some attempts were quickly discontinued because of the deluge of litigations, fights and complaints of all kinds, when trying to check the existing records. One must also add an objective difficulty : land consolidation is particularly difficult in areas where the land is both low (flooded or submerged) and high. It would be ideal to simultaneously conduct fool proof drainage and consolidation operations. Otherwise a farmer may rightly prefer to keep scattered plots, some flood prone, others bearing no such risk.

All these shortcomings in planning and management are closely connected with Bihar politics, thus involving two sets of interrelated problems:

a) the lack of real political leadership over the last two decades if not more;

[8] Reserve Bank of India, *op. cit.*, p. 301. Shifting courses of the rivers create additional difficulties: semi-circular *chaurs* (water-logged areas) in different directions.

[9] Reserve Bank of India, *op. cit.*, pp. 298, 299. The very widely advertised Training and Visit System of extension promoted by the World Bank has not been successful, as could be expected. See some references in my *India's Changing Rural Scene* (New Delhi, Oxford University Press, 1982).

b) acute caste and class struggles, involving the high castes, rising middle castes/classes, and Scheduled Castes in complex networks of rivalries.

Regardless of their party affiliations, state governments have been known to be unstable, weakened by the calibre of the politicians and corruption. Though one comes across efficient and dedicated senior civil servants in Patna and in the districts, their capability is too often curtailed by politics.

Caste and class struggles have worsened during the past fifteen years. In 1978, one heard fewer instances of violent battles than one does now.[10] Such incidents have become widespread in the plains of south Bihar, Patna, Gaya and Nalanda districts for instance. Although they are not common north of the Ganges, the situation is not particularly peaceful as seen in Kalpi.

Finally, how should we weigh the impact of human behaviour: 'The ease loving complacent attitude of the people, allergic to hard work', quoted in the old Gazetteer of north Bihar, for example? Such a judgement can be heard even today. How far is it correct? I am uneasy about such sweeping statements. As seen above, many complex, techno-economic and socio-political factors are involved. While we cannot entirely reject these arguments, we should also consider the time factor. True the taboo on ploughing is still alive in Kalpi, unlike in Nilkanth, but it may be gradually eroded. Some formerly unenterprising Bhumihars today show more interest in developing their land. In other districts, as in parts of south Bihar, this trend is stronger.

In the face of these complex issues, two points come out clearly: the expansion of the economy remains too slow to really affect large numbers of small farmers and landless labourers. Second, the magnitude of water management problems is frequently overlooked by people outside Bihar.[11]

[10] The use of sten guns by retinues of landlords was considered a great, but worrying novelty in 1978. It is common now.

[11] A point emphasised by one of the most experienced senior officers from Bihar, during my lecture at the Centre for Policy Research, New Delhi, February 1986.

Chapter Nine

Similar Water Management Problems in a Different Socio-economic Environment

Let us proceed to another potentially rich but sluggish area, Puri district in the delta of the Mahanadi in Orissa.[1] Water management problems here are similar to those observed in the middle Ganges basin, but within a historical background and socio-economic environment very different from that found in Varanasi and Muzaffarpur.

Orissa must have known a fairly high level of economic activity. This is evident from the numerous temples dating back to the middle ages, especially those built between the 7th and the 13th century. Orissa has some of the largest temples in the subcontinent — the Lingaraja at Bhubaneshwar, Jagannath at Puri and the Sun temple at Konarak.

The sequence of later events is not too clear — at least for me — but the area was fairly underdeveloped when the British took over. In addition, Orissa remained one of the most neglected parts of the Empire, very isolated, with no industry and poor infrastructure. The coastal districts became an autonomous province only in 1936. After Independence, several small princely states were merged into the province.

When I first visited Orissa in 1954, even the highways were not asphalted. Life was traditional and due to a slower population growth and relatively more space than in the Ganges basin, agricultural production was able to meet the requirements of the province, which even enjoyed a small surplus of grain. Within the next thirty years, however, in spite of a definite move towards development and modernisation, Orissa faced a deficit in foodgrain.

[1] Part of the district outside the delta has not been surveyed.

Poor people and poor agriculture

Instead of focusing on a single village, I decided to work in two different blocks, visiting several villages and hamlets.[2] Let us start in a particularly poor area: Brahmagiri block.

The overall agricultural situation has not changed much since my first visit. Out of a net cultivated area of 19,000 ha. in the block, 10,700 ha. are low-lying land which suffer practically every year from excess water. Paddy is sown broadcast and grows with the rising water and with plenty of weeds. It is of course of no use to try new seeds and chemical fertilisers. No wonder, then, that paddy yields remain around 500 to 800 kg./ha., as has been the case for several decades. On intermediate and high land, yields range between 1,000 and 1,500 kg./ha., following the rains. Some farmers may be lucky enough to harvest a little more.

In the *rabi* season, the moisture left in the soil enables farmers to now sow about 11,000 ha. (half with paddy and half with pulses), an area which has expanded here, as in many other parts of Orissa. In the past ten years, *rabi* crops have benefited from irrigation. Lift pumps have been installed on a major drainage canal, so that 320 ha. of vegetables, potato and chillie are now irrigated.

Though such progress could be considerably expanded, water management during the monsoon remains as inefficient as ever. Vast areas are submersed for lack of drainage. Low-lying land near the coast is also exposed to salt infiltration at high tide.

Even with a relatively low density (270 per sq.km.in 1981), the people (who totalled 84,888 in 1981) lead a difficult life.

Twenty kilometres east of Puri, we leave the good district road and move towards the sea to Chandrapur village. The old sandy track has been replaced by a narrow asphalted road. Poor paddy fields had been harvested, or were about to be (November). Small hamlets can be found clustered under trees and fringes of casuarina woods extend towards the sea.

We come first to the houses of the Dhoms (a Scheduled Caste). Govin Mallik, whom I had interviewed in 1978 had passed away. In those days, he and his family made bamboo baskets which brought them Rs. 2 to 3 per day. They also worked as field labourers

[2] I made only two surveys in 1978 and 1985 after several visits to Puri district without any real research since 1954.

earning Rs. 3 to 3.50 a day without food. Today, Govin's son seems to be better off, like his neighbour, Hari Mallik. The latter has been given 3 acres of land recovered from surplus land by the government. True, the value of the land is not high, but they have started growing cashew trees and some casuarina. Around both houses, papaya trees have begun to bear fruit. They also have a few chickens. Basket making is on the increase. Both families buy bamboo worth Rs. 6 to 7, enough to make four baskets. These are sold at Rs. 5 each and are in great demand as they are used for offerings in the temple of Jagannath in Puri. One of Hari's sons has been able to buy himself a watch. He has one cycle like his neighbour, who also owns a transistor. They have also been given a subsidy of Rs. 1,600 to rebuild their houses. The walls are made partly of brick and partly of mud and they have corrugated iron roofs which offer better protection against the rains than do thatched roofs. One of Hari's sons also trades in goats and this brings him a profit of Rs. 200 to 300 per month. He started this business with a government loan-cum-subsidy of Rs. 3,000 (Rs. 2,000 as loan and Rs. 1,000 as subsidy). His brother works in Puri as a rickshaw puller. They all have three sets of clothes now, instead of only one set each as in 1978.

In addition to these various activities, all the men work as casual labourers during the paddy season earning Rs. 7 per day either for transplanting or harvesting. It was heartening to find such progress in such an isolated place. The diversification of activities, influenced by the needs felt in Puri, has more than compensated for the lack of increase in agricultural production.

We proceed further, past the Muslim quarters, to revisit the hamlet of the Dhobis, Here the situation remains quite depressing. Boni Sethi lives no better now than during our first visit. The other Dhobis are also in a similar condition. Earlier, Boni told us, he would receive 170 kg. of rice whereas now he gets only 120 or 130 kg. per year as payment for washing clothes for the higher castes. He also received 1 ha. of land from surplus land, recovered by the government but the soil is sandy and the land undulating. He has some casuarina trees and has started growing cashew trees as well. For three months he is busy transplanting, weeding and harvesting paddy at the rate of Rs. 7 per day without food. His wife, no longer alive, used to be responsible only for washing clothes. . .

Lakshman Sethi has a tiny plot of land (0.2 ha.) which gives him less than a quintal of paddy. He has now been married for four or five years and his first two children died soon after birth. Luckily he owns a bullock which enables him to cultivate 0.6 ha. on a sharecropping basis, which for him means 100 kg.of paddy. He also works as an agricultural labourer.

Around us, several children are totally naked. The huts are almost bare except for a few cooking pots and mats. In one of them an old woman lies groaning and it looks like the end is near for her.

In the Muslim quarters, we met one of the relatively rich men who owns 8 ha. of land : 4.8 ha. are planted with cashew trees which should bring him a high income, 3.2 ha. of paddy yields 750 to 1,000 kg./ha., except a few plots located at a higher level which yield 1,250 kg./ha.

One cannot talk of a fall in the standard of living. Between 1978 and 1985, daily wages have at least doubled from Rs. 3 or 3.50 to Rs. 7, while the retail price of 1 kg. of rice has risen from Rs. 1.75 or 2.0 to Rs 2.40 – 2.60 (just after the harvest, Rs. 2 to 2.20). Yet, left to themselves, the weaker sections of society remain very poor. The Dhoms were lucky to get some government help. They were also better able to find other jobs (basket making) than were the Dhobis.

Obviously, to reach a higher level of living, people, and not only the poorest, need to improve farming conditions on a much larger scale. One way to improve the situation would be to push irrigation in the *rabi.* Now, out of 19,000 net cultivated hectares, only 800 are irrigated. The progress of lift-pumps could also be accelerated as it would take much longer to improve monsoon crops through proper drainage.

Pipli block

The block is located on the main road leading from Bhubaneshwar, the new capital of Orissa built after Independence, to Puri, one of the holiest cities in India. When I first came in 1954, the non-asphalted road carried little traffic, whereas now, buses, trucks and bullock carts are numerous. Viliages by the road side are expanding, tea stalls for the pilgrims, shops with plastic goods and local handicrafts are flourishing. The city itself is growing. In addition to the rising number of pilgrims, we now find buses loaded with Indian tourists. Gone are the days when one could wander around the Sun temple at Konarak practically alone, as I did in the 1950s and the 1960s.

Already in 1978, one could feel the impact of the road and of the holy city on the villages nearby. The overall level of agriculture was a little above that of Brahmagiri block. The area being less isolated, the wages of agricultural labourers were a little higher : Rs. 3.50 to 4.00 as against Rs. 3.00 to 3.50. Some small farmers were developing kitchen gardens, selling their vegetables along with coconuts from recently planted trees. However, in those days wages were declining in real terms, but employment opportunities might have risen slightly, even if at low wages.

	1971	*1981*
Population	84,000	115,360
Density per sq.km.	404	557

(Geographical area 207 sq.km.)

The increase during the past decade appears extremely high. Has there been some underestimation in the 1971 Census ? At any rate, average density is becoming heavier.

Pipli block is located in the command area of the irrigation canals coming from the Hirakud dam, upstream on the Mahanadi. Although the dam was inaugurated in 1957 and the main canal reached Pipli in 1967, our first survey (1978) showed that the irrigation potential created was far from fully utilised, a typical instance of long delays common to many major irrigation works. In several areas, field channels were missing, so that water could not be properly utilised. After long debate and discussion, it was finally decided that the state would build such channels instead of the farmers, the first of which we saw in 1978.

Not only was irrigation not fully utilised but the entire command area suffered from the lack of proper drainage. Originally it had been assumed that the existing drains would be enough, but this was proved wrong. Slowly, new drains were being built although wide tracts of land were still at the mercy of excess water due to floods and to the heavy rainfall (around 1,500 mm.) during the monsoon.

Total net cultivated area	19,400 ha.
rabi	12,470 ha.
kharif	15,570 ha.

While these data have not changed much in the past ten years, the irrigated area has increased.

	1977-78	*1984-85*
Irrigated area : *rabi*	5,600 ha.	10,370 ha.
kharif	9,000 ha.	9,000 ha.

Paddy was and remains the dominant crop, covering most of the area in the *kharif* and about half in the *rabi,* the remaining half bearing pulses, vegetables and potato.

The average yields of paddy for the block show considerable differences. Vast tracts of land, particularly just before entering Puri, are submersed for weeks in the monsoon. There, 830 ha. can have 1.50 to 2.00 m. of water. The local broadcast variety manages to survive but with a yield of 500-800 kg./ha. On other low-lying land the yields of broadcast paddy may reach 1,000 kg./ha., a situation which is hardly surprising when one considers how slowly drainage is progressing. In 1978 and again in 1985, one came across vast areas without drains. Elsewhere, drains were choked with grass and were of little use. Besides, the rising water table due to the seepage from irrigation canals created waterlogging in certain areas. Transplanted paddy on intermediate and high land used to yield 1,200 kg./ha., unirrigated, and 1,750 irrigated. In the past few years the yields have risen to 1,700 and 2,500 to 3,000 respectively. New seeds are used on irrigated plots along with small doses of chemical fertilisers.

In the *rabi,* new seeds and chemical fertilisers often yield (under irrigation) 3,000 to 3,500 kg./ha. As will be seen later, some farmers are now able to reach 3,000 to 4,500 kg./ha. Sometimes, though rarely, 5,000 kg./ha.

Unlike drainage, irrigation has expanded. Now 110 villages out of 217 have a network of field channels, which explains the recent increase in irrigated area. The government is also encouraging the construction of tubewells (public and private) which are particularly needed in the *rabi* for vegetables, chilli and potato. Eighty-four units have been installed so far..

While far from spectacular, paddy production is rising in spite of serious constraints like poor drainage and incomplete irrigation. Coconut trees and vegetable production are also progressing, as well as transportation and trade. These combined factors explain the trends in wages

Daily Agricultural Wages (Rs.)

1978	*1985*
3.50 – 4.00	7.00 – 8.00

Rarely, wages may rise to Rs. 10. The trend in real terms, though not spectacular, is nevertheless typical of a moderate growth of the overall economy.

In Mangarajpur village, I interviewed a group of landless labourers. They are busy with paddy cultivation for sixty days during the monsoon. In the *rabi* they have four months of work attending vegetables and another difficult job — tending betel leaves.[3] Each labourer owns two or three sets of clothes. As a result of being able to find additional jobs, they are now employed for ten to eleven months in the year rather than for eight months, as was the case ten years ago. Their wages amount to Rs. 8 per day from agriculture and Rs. 7 from tending betel leaves, a job which takes two hours a day. They own their own homesteads.[4]

Near the main road we stop at the houses of some Harijans. Rabindra Malik lives with his two brothers. Together they own 0.4 ha. yielding 750 kg. (1,875 per ha.) of paddy, followed by 120 kg. of pulses. They have one bullock. During the monsoon they find work as agricultural labourers for Rs. 7 per day or go fishing, earning Rs. 10 per day. Around their own homestead they have twenty-five coconut trees[5] producing eighty nuts per tree, sold at

[3] Vines grow under a shade. Great care must be taken to ensure adequate moisture.

[4] In many villages poor people have been allotted ownership rights on their homesteads, a not insignificant step which frees them from pressure from the landowners.

[5] For a long time, coconut trees belonged mostly to Brahmins. When other castes began growing them, the Brahmins tried to keep their monopoly by warning lower castes that it would be inauspicious for them to grow such trees and that if they did they would die!

Rs. 1.50 per piece. When I asked them if they are better off now they replied in the affirmative. They are quite smart, have a few clothes and wear sandals. In the evening they listen to their radio.

When revisiting a hamlet surveyed in 1978 where some families of Dhobis live, I found that one of them now owns 0.5 ha. where, thanks to new seeds and fertilisers, they harvest 2,000 kg.of paddy followed by pulses.

In Kamjra village, landless labourers earn only Rs. 7 per day. Local people complain of the competition from people coming from Brahmagiri which depresses wages. The Chamar whom I interviewed is nevertheless not too poor. In addition to agricultural work, he makes and sells bamboo baskets. Around his homestead graze two or three goats which he will sell after one or two years. He admits that there has been some progress in his life : he now owns a cycle and two sets of clothes.

D.K. Sahu belongs to one of the main local high castes, the Khandayat. With his three brothers they own 0.8 ha. In a good year, they harvest the equivalent of 2,815 kg./ha. of paddy, a high yield for the block. Yet, twice this year the nurseries were washed away by torrential rains. They had to sow again, and when I met them, they were not expecting more than 2,250 kg./ha.

Vasant Kumar, a Brahmin, owns 4 ha. of land with his four brothers. Half is low-lying land with a very poor yield, the other half yields 1,125-1300 kg./ha. Vasant Kumar looks after the land. Two brothers work outside the village, one as a clerk, the other as a teacher. The fourth brother sits idle at home. When we met him, Vasant Kumar was supervising some workers who were digging a pond where he intends to breed fish. The earth dug out will create elevated land which will be suitable for vegetables and banana and will be irrigated from the pond. He is quite clever about improving his land. Yet, when I asked him if Brahmins handle the plough here, he replied: 'oh no, we would be outcaste.'Here then is a case of a family which keeps on increasing along with the division of their holding. Yet, belonging to a high caste and being more educated, they are able to find jobs outside their village more easily.

Govind Panda, Khandayat, has 1 ha. of land. He has only one son who is at school and one daughter who is married. He owns a pair of bullocks and has recently installed a filter point (shallow tubewell) which cost him Rs. 18,000, half covered by a loan, half by a subsidy. He has extra water to sell to his neighbours, from which he earns Rs. 600 per year. The land is never submersed so that, with the help of new seeds and chemical fertilisers, he harvests 3,700 kg./ha. during the *kharif* season. Just before the harvest he sows black gram (pulses) by throwing in the seeds along with the paddy. The harvest is at the beginning of February. His filter point gives him another paddy crop until May. Not only does Govind enjoy a high intensity of cropping, but because of his pulses which fix the nitrogen from the air, he is able to enrich his land at a low cost. Close by one comes across another filter point where paddy is followed after the monsoon by vegetables, potato and groundnut, so that three crops can be grown a year.

One could not find better examples of the potential of the area. Drainage, if necessary, combined with perennial irrigation (which can depend on both canals and tubewells), permits the use of high yielding varieties, multi-cropping and a diversity of crops. This ensures that farmers with 1 ha. of land or even less are able to live quite well.

Unfortunately, one must again underline that even now, such progress remains fragile. Farmers complain of frequent power cuts. They may also be compelled to give a *baksheesh* to the officials of the electricity board as a means of ensuring their supply of elctricity.

Parts of the block with intermediate or high land with canal irrigation, provide us with other cases of families improving their condition of living. Madu Sadan Sahu, a Khandayat, owns 2.4 ha. His whole family, including his father and mother and three married brothers constitutes eight adults and three children. Earlier they were unable to produce enough rice for their own consumption. The men used to find casual work in order to earn some extra income. Now, 1.2 irrigated hectares yield the equivalent of 4,700 kg./ha. and unirrigated plots yield 2,250 kg./ha. The family is now able to sell some surplus paddy and for the past one year they have been running a small rice mill. They also cultivate betel leaves which are used for *paan*. Several other small farmers in the same village produce from 3,000 to 5,000 kg./ha. of paddy per crop.

A glimpse of the district

The district[6] has plain alluvial tracts (described above) and hilly

[6] The following data have been supplied by the district authorities. See also Puri *District Statistical Handbook 1974* and *1975*, Puri District Statistical Office, 1977.

tracts in the interior. That is why average district data are of little use. For instance, the average density for the district is 287 inhabitants per sq.km., which obscures the deep differences within the district. This is evident from the fact that the density in Pipli block is 557.

Total geograpnical area	10,182 sq.km.
Population	2.34 million in 1971
	2.92 million in 1981

About one-third of the area cultivated in the *kharif* (150,000 ha.) is irrigated, mostly by canals and about the same ratio is irrigated in the *rabi* (100,000 ha.). Tubewells and filter points with 825 units still play a very minor role in spite of the large potential of ground water. Paddy is by far the dominant crop:

1984-85

	Total cultivated area (ha.)	Irrigated area (ha.)
kharif	355,000	130,000
rabi	32,000	30,000

About 170,000 ha. of paddy are classified as low-lying land with a minimum standing water of 15 cm. (the ideal situation), but not infrequently up to 1.5 m. Half the low-lying land is often severely affected by excess water. Sometimes the whole crop is destroyed, sometimes yields fall to 500 kg./ha. of paddy. Intermediate land (140,000 ha.) is rarely submersed and, especially with some amount of irrigation, can, as seen earlier, give a good yield. As for elevated land (40,000 to 80,000 ha.), often found in the interior, half the area suffers from a lack of water because of a high run-off in hilly, red soils (see next chapter).

Pulses (grown mostly after the monsoon) have expanded from 160,000 ha. in 1978 to 220,000 in 1984-85, yielding 350 to 500 kg./ha. Vegetables have more than doubled in the area between 1973-74 and 1984-85 to reach 52,000 ha.

It is not easy to make a clear cut assessment of the achievements during the past ten years. Irrigation is progressing, but relatively slowly. The magnitude of the problem of drainage too does not seem to have been reduced between our two surveys.

New seeds of paddy and chemical fertilisers have begun to be felt. The consumption of chemical fertilisers has risen from 7000 t.

(nutrients) in 1977-78 to 11,338 t. in 1983-84 and, 11,104 t. in 1984-85. However, this is still quite a low level for a total gross cultivated area of about 700,000 ha.

In the 1970s, Puri district was a district deficient in rice. Now it enjoys a small surplus. This is, to a fairly large extent, due to the extension services. The deputy director of agriculture, in charge since 1978, is a man of outstanding calibre, probably the best agricultural district officer I have met in India. He is well qualified in agronomy, has a great deal of experience and energy. His staff is equally efficient. He is very often in the villages, talks in a friendly manner with the farmers and goes out into the muddy paddy fields to check the number of hills of paddy himself. His enthusiasm is infectious. Under such a person, the Training and Visit System of extension introduced with the help of the World Bank does yield good results. The question is to know what matters more : the system per se, or the head of the operations, a question to which I will revert later in this book.

On the other hand, even such efficient extension services are constrained by the shortcomings of the water management system, a sector under the state irrigation department.

The production of vegetables, coconut and chilli is expanding, partly as a result of the growing needs of Puri. Trade and transportation are also important. This is confirmed by the relatively fast rate of urbanisation which has risen from 10.8 per cent of the district population in 1971, to 17.3 per cent in 1981.

We do not have enough data to assess the trends of the labour force in terms of occupation and the profile of landholdings. Data for all the districts of coastal Orissa give an average density of 310 per sq.km. (1981). The work force amounts to 10.7 million of which 45 per cent are cultivators and 26 per cent are agricultural labourers. Owners of holdings below 1 ha. represent 53 per cent of the total landholders. Another 31 per cent own between 1 and 2 ha.[7] This break up of operational landholdings implies a dangerous and rising population pressure on land, although the percentage of labourers is not yet as high as in Bihar.

It brings us to conclusions similar to those in our observations in the middle Ganges basin : there is an urgent need to drastically

[7] Reserve Bank of India, *Report of the Committee on Agricultural Productivity in Eastern India*, vol. II, p. 182.

improve water management, which would pave the way for higher yields and more multi-cropping.[8] At the same time,the process of diversification of the rural economy needs to gain much more ground. It is also encouraging to see the rising percentage of urban population, although this trend needs to gather still more momentum in order to reduce the population pressure on agricultural land.

While the economic situation presents a more positive picture here than in north Bihar, the socio-political climate is also more healthy. One hears, no doubt, complaints from the poor of exploitation or pressure by the landowners, but one does not come across the violent and bloody clashes and class/caste conflicts found in Bihar and less so in eastern UP. The local caste structure is different. Unlike in Bihar, for instance, one does not come across big landlords with an armed retinue. This of course does not preclude more tensions in the future.

While on the one hand agricultural services operate well, one hears complaints about various malpractices. Again, I have formed a mixed picture of the DRDA and its programmes for the poor I have encountered some classical stories of *baksheesh* given for a loan-cum-subsidy, but also the remarkable project for fishermen, who with the help of a loan-cum-subsidy, were able to buy boats and nets and increase their catch.

To sum up, the coastal areas of Puri district are no longer dormant. Technical innovations are spreading ; there is some diversification of the economy, more vegetable production and coconut trees, more trade and transport, all of which is encouraged by the holy city, the larger number of pilgrims, the growing number of Indian — and to a lesser extent of foreign — tourists visiting the area.

Some poor people have known an improvement in their condition of living, and wages, after having stagnated in real terms for many years, have now risen. However, the whole process of development needs to be encouraged more vigorously in order to bring about sizable and definite social progress.

[8] It will be interesting to watch the impact of per cent land consolidation on irrigation. It started in 1974 in Orissa. By now 60 per cent of the cultivated area in Puri has been covered. Usually landowners do not obtain one single plot of land. Normally they get one of low-lying land, one intermediate land.

Chapter Ten

Rainfed Paddy in the Plateau Zones

It is time to take leave of the alluvial plains which have a large irrigation potential, to enter the vast peninsular zone of the Deccan, where often soils are poor and in the process of degradation. The monsoon varies considerably from high rainfall in the north-east, to semi-arid conditions in the central parts, and arid conditions in western Rajasthan. Then, in many areas rains cannot be complemented by irrigation due to insufficient surface and ground-water potential. No doubt in such regions the population pressure is not as heavy as in the alluvial plains, but yields of crops are also much lower and multi-cropping less widespread.

In Deoghar district [1]

The plains south of the Ganges with their vast tracts of paddy fields come to an end towards the Chotanagpur plateau. Low and rocky hills and valleys gradually give way to mountains that rise out of the horizon, covered with forests, some old and dense, others turning into bushes following the indiscriminate felling of trees.

Santal Parganas district was, like several other districts in Bihar,[2] divided into four districts in 1983, an important and welcome measure because the former district was much too vast to be adequately administered. Deoghar, one of the new districts, covers 2,451 sq.km. with a population of 708,828 in 1981. The density per sq. km. is 290, a relatively high figure considering the available resources.

From Deoghar town, we proceed to the village of Lerwa where several families of Paharias live: an adivasi tribe more or less

[1] I made only one survey in November 1985.

[2] Districts inherited from the British Raj were very large.

Hinduised, as are many other Scheduled Tribes in the area.[3] A voluntary organisation[4] (Santhal Paharia Seva Mandal) provides assistance to the poorest people through loans and subsidies, especially for digging wells and for housing. One farmer, for example, owns 1 ha. of land; on 0.8 ha. he grows rainfed paddy, yielding around 1,000 kg./ha. or a little more; in the *rabi* he grows wheat on a small plot. He also grows some pulses like red gram. With the help of a loan he now has a well which irrigates 0.2 ha. and a well-kept kitchen garden from which he sells vegetables and potato. The Mandal is helping farmers to improve the existing tank in order to expand irrigated wheat in the dry season after paddy. They are also helping people to build better houses.

On the whole, these people are not too poor, an impression confirmed in the next village, Sujani. Here we find Santals, the most important Scheduled Tribe in the region. They too live off paddy and are now growing more wheat with the help of a tank and a low-lift pump. This is only the beginning and yields are still not more than 900 kg./ha. Most people do not have enough land to cover even their basic needs. For about two months they find work in paddy cultivation, earning Rs. 10 to 12 and some food. Most of them have at least two sets of clothes, though semi-durable goods, like bicycles are rare. Wages are quite high, which can be attributed to the proximity of Deoghar town.

For 20 km. towards the north-east, we go through increasingly deforested hills, some entirely barren and rocky, others covered with bushes. Banjhi village and the surrounding area were carved out of the forest half a century ago. Santals and Chamars both live here. The latter work for the former. For transplanting paddy they receive 3 kg. of grain along with two meals. At the time of harvest they get around 5 kg. of paddy[5] and one meal. These wages have improved compared to what they were twenty years ago, when for transplanting they would receive 1.5 kg. and for harvesting, 3.6 kg. Yields vary from 1,000 down to 500 kg./ha. The Chamars also make *beedis* (local cigarettes) for which they earn

[3] Deeper into the interior, often in the forest areas, one comes across tribals who are animists.

[4] Created in 1941 to fight leprosy, the Mandal is now engaged in several fields: health, education, agricultural extension, handicrafts, etc.

[5] Retail price of 1 kg. of paddy: Rs. 1.60.
Retail price of 1 kg. of rice: Rs. 3.50.

Rs. 6.50 per day, a woman's job lasting two or three months. The government has assisted three farmers in digging wells, each of which irrigates 0.8 – 1.2 ha. with the help of a pump. Eight hand pumps have also been recently installed.

In the entire village, 50 ha. are devoted to paddy which cannot be irrigated by wells because of the lack of a sufficient supply of water. Twenty-four hectares bear *kulti* or horse gram (pulse) harvested in November : 650 to 750 kg./ha., sold at Rs. 2 per kg. In the dry season 5 ha. bear irrigated wheat. The wells are used mostly for kitchen gardens where vegetables, chilli and potato are grown.

Potilal, a Santal, owns 0.4 ha. where he cultivates paddy. As a labourer he earns Rs. 6 and some food for transplanting. For harvesting he receives either 5 kg. of paddy per day or Rs. 7 but no food. Ten years ago transplanting brought 1.5 kg. of grain or Rs.2 and harvesting Rs. 2.50 or 2 kg. One can see that there has been some increase in real terms. Besides, Potilal has a cow which gives a little milk. He had spent some time in the home guard and was able to save some money to buy a watch. Quite lively, Potilal seems to manage fairly well.

Budhan Kishu lives with his wife, his two sons and two daughters. One of them returned to her father's home after divorcing her husband. The agricultural holding covers 3.2 ha. They own four cows and six bullocks. Two ha. are devoted to paddy, 0.8 ha. to horse gram and the remaining plots to vegetables and chilli. Budhan also has chickens, pigeons and goats which he sells from time to time. He draws some benefit from the forest where his animals graze.

Latasary village is rather similar. Paddy remains the major crop, followed by horse gram and other pulses. The few wells installed irrigate nice kitchen gardens which supply vegetables for sale. We met some Chamars who raise pigs, others who make *beedis* or shoes for additional sources of income. All these people seem less poor than the Chamars of Kalpi. One also does not feel social tensions as acutely. They agreed that their lot had improved slightly.

The proximity of the town to our first group of villages, the resources drawn from the forest (often illegally), the expansion of ordinary wells and subsidiary activities may explain why one did not come across very acute poverty. Besides, in the other areas visited, one could feel the impact of the Mandal and the results of government schemes for irrigation, which are not to be found in all villages.

At the district headquarters, we benefited from good and elaborate briefings on the main local problems, meeting several officers, including the district magistrate.

Net cultivated area	95,520 ha.
Current fallow land (during one year)	36,610
Paddy	75,160
Wheat	5,930
Maize	6,960
Potato	1,100

Exclusively rainfed paddy can yield up to 1,200-1,500 kg./ha., but if the crucial *hathia* rains (end of September) fail or are delayed the crop will suffer. Especially on elevated lands where moisture is less, losses can amount to 20 per cent or more. Neither new seeds nor chemical fertilisers are used.

Irrigated wheat, a rather recent development, reaches 1,200 kg./ha. in the best cases, but more often stands at 800 – 1,000 kg./ha. It could go up to 2,500 kg./ha. in well irrigated fields. Pulses give 600-700 kg./ha., a fairly good yield. So far the forest (21,200 ha.) has been an alternative resource to fall back upon in terms of grazing and cutting of wood, both practised illegally in many cases. Now with an average district density of 290 per sq. km., additional resources are badly needed.

The authorities are pushing irrigation which now covers 9,900 net ha. i.e., one-tenth of the cultivated area. Double-cropping amounts to only 4,600 ha. Tanks are the major source of irrigation. In addition to small private ones, the government has constructed 182 tanks, each irrigating around 20 ha. The trouble is that they are of course vulnerable to drought. Besides, soils are often porous so that much water is lost.

Since 1978-79, twenty-five schemes of low-lift pumps (25-50 HP) have been completed, each with a command area of 120 ha. in the *kharif* and 80 ha. in the *rabi*. Most of them are electrical pumps and enjoy a relatively reliable supply of electricity from the Damodar Valley Corporation (15-20 hours per day). Each project has cost Rs. 500,000 (1980 prices) and requires Rs. 30,000 per year for maintenance and operations. A boring is made in a river bed, seven metres deep, and the pump is installed on the bank. Water is usually available for most of the year.

Such interesting schemes unfortunately face several constraints. The first limitation is the lack of water, which prevents a massive expansion of irrigation, bearing in mind the present, very low level of irrigation. The second problem is that of electricity. The irrigation department could start another hundred projects, but they are not sure of getting enough electricity. Third, both tanks and low lift pumps will require a strong and efficient organisation for maintenance and operations.

Private irrigation relies on sixty tubewells, 455 ordinary wells (as seen in these villages) constructed on government subsidies besides other private wells and ponds.

Drinking water is another serious problem. A new model of the hand pump is being installed in villages, often through a loan-cum-subsidy, but in such a rocky terrain the investment amounts to Rs. 12,000.

Under such conditions, the future of agriculture is far from bright. Rainfed paddy may not increase much above the present yields of 1,000-1,500 kg./ha. The net area of irrigated wheat is bound to remain small. So far one does not see much progress in pulses. Soil erosion is increasing, which requires a number of complex and expensive measures as well as research. Here again, one cannot expect instant remedies. Finally, the rather high normal rainfall (1,300 mm.) is misleading. During our visit in 1985-86, the rains had been good. In 1984-85, the total rainfall looked adequate, but the *hathia* rains had failed, which affected the crops. The two previous years were fair but 1981-82 faced a very severe drought.

What are the other possible resources ? Forests are already over-exploited, although more efforts are being made for reafforestation. Accacia plantations look promising, both under private initiative and under the forest department. Yet, the protection of forests remains an irksome issue. Farmers encroach upon the forest, collect wood illegally, while forest department officials indulge in various malpractices.

Out of 100,000 agricultural holdings, one-third are below 1 ha., 25,000 between 1 and 2 ha., 14,400 between 2 and 4 ha. and 27,000 above 4 ha. Taking into account crop conditions, the owner of 2 ha. of land will have enough food grain for his family but very little surplus, if any, for sales.

As is usual in peninsular India, the percentage of agricultural labourers compared to all agricultural workers is relatively lower here than in the plains of east and south India.

District occupational distribution (1981)

Cultivators	117,843
Agricultural labourers	32,394
Allied agricultural activities	8,587
Cottage industries	9,754
Other industries	7,674
Trade	4,351
Others	28,944
Total workers	208,717

Thus, 76 per cent of the total population works in agriculture. The urban population is also, and not surprisingly, low : 12.6 per cent of the total population of the district.

In our brief tour of the villages, we did not see extreme cases of poverty, but we did see areas which had benefited from several projects aided by the Mandal and the government. Besides, the first village was near Deoghar town and this accounted for its higher labour wages compared to more distant villages, a pattern very common in India and many other developing countries.

Daily wages for paddy cultivators

	1975	*1985*
Transplanting	1.5 kg. or Rs. 2	2.3 kg. or Rs. 5-6
Harvesting	2 kg. or Rs. 2.50	3-5 kg. or Rs. 7

Near Deoghar wages rise up to Rs.10. All wages are supplemented with some food. The data for 1975 have been given by the farmers and I cannot guarantee their accuracy.

There are other areas in the district which are much poorer, due to more limited resources, greater isolation and other factors. Many Paharias face increasingly deteriorating conditions. According to the 1931 Census, they numbered more than 200,000 as against 100,000 in 1981 : 'This tribe is continuously diminishing due to [the] unhygienic climate and non-availability of clean water and adequate food,' wrote the Mandal to me later on. In a news conference, the Bihar welfare minister confirmed that the Paharia tribal community was on the verge of extinction.

Once again we come across the perplexing question of special schemes for the poor. The very dedicated and efficient team of the Mandal does a fine job but like any voluntary private organisation, it remains unavoidably confined to small areas. The government schemes which we visited also made a good impression, but others are not as successful. In his press conference, the Bihar minister for rural development was highly critical of the IRDP. In five years only 339 Paharia families were rehabilitated. Many cases of malpractices were revealed : bullocks and goats supplied on a loan-cum-subsidy basis at double the market price, goats in such poor conditions that they quickly died, etc.[6] The IRDP also suffers from the problem of very poor loan repayment.

Deoghar district is a clear case of an area which needs massive economic diversification, in view of the relatively limited scope for agricultural development. A new industrial town has been created at Jasidih, near Deoghar. Since 1980, forty-eight factories have been set up, employing 1,500 workers. They enjoy incentives in terms of subsidised loans and power supply. Deoghar itself attracts many pilgrims who walk from the Ganges up to the famous temple of the city, which creates some employment. However, such trends have carried limited weight so far. Besides, the whole area is rather isolated from any major industrial centre.

I could not gather information on migrations outside the district, towards the mining and industrial areas in the Damodar valley or Jamshedpur. There may be some outlet there, but many poor people from other parts of Bihar also migrate there.

Keonjhar district

After a very brief visit in March 1984, I came to Keonjhar again in November 1985. For many kilometres one drives through hills which give way to superb mountains covered with forests. In the midst of them appears Jamshedpur, the steel and engineering goods city built by the Tatas, where India is at its best in terms of efficiency, social welfare, modernisation and expansion. And yet, no more than 30 km. away roam wild elephants.

Keonjhar (8,303 sq. km.) lies further, in the northern plateau of Orissa. Nearby, half the area is covered by forests and mountains

[6] *Overseas Hindustan Times*, 24 May, 1986.

reaching above 1,000 m. In between are undulating plateaus with small terraced paddy fields depending on the rains. Around 45 per cent of the population are tribals. Some of them are very isolated in the mountains and still follow the old technique of shifting and burning cultivation, using only the hoe. They maintain their other traditions and customs which remained aloof from the influence of the main currents of history. Other tribes are more or less Hinduised and are in closer contact with the outside world. They use the plough and bullocks.

Keonjhar was a princely state until its merger with Orissa in 1948. The whole area was little touched by modernisation until the beginning of the five-year plans. The economy progressed slowly, while the population began to increase faster : 7,43,315 inhabitants in 1961, 1.1 million in 1981. In spite of its low density (133 per sq. km.), the district is deficient in foodgrain. The average rainfall is quite high (1,600 mm.) but as will be seen, erratic from year to year. A wide plain surrounds the small town of Keonjhar towards the south. The harvest of rainfed paddy, in full swing during our visit, looked good with yields around 1,500 kg./ha. In a drought year it can fall to 500 kg./ha. On a few plots they have tried new varieties with chemical fertilisers, yielding this year (1985) 2,500 kg./ha. A few kilometres away from the city, wages for transplanting paddy amount to Rs. 6 or 7 and for harvesting, to Rs. 8, in both cases without food. Groundnut cultivation is encouraged, giving fairly good yields : 1,500 kg./ha. unshelled, sold at Rs. 500 per quintal. It is obviously much more profitable than paddy, but groundnut can be cultivated only on relatively high land for fear of floods.

Banko Patuli is a village 15 km. from Keonjhar. We first come across Dai Tari Prushti, a Banya (trader) by caste, but involved in a number of activities. His father had no land, earning his living by crushing oilseeds. He himself now owns 1.6 ha. of land devoted to paddy out of which he gets 2,000 kg. He has four cows and two bullocks. One of his sons has become a contractor employing ten workers. All of them live in a well kept house. Their well irrigates a good kitchen garden, surrounded with a few papaya and coconut trees. Our hosts are smart and lively. The young contractor was about to leave for Keonjhar on his motor-cycle: an evident success story where entrepreneurship played a major role aided also by the relatively high status of the family in terms of the caste structure.

Other villagers do not live as well. In addition to paddy, they grow some pulses (*arhar* or red gram) in order to earn some money : Rs. 400-500 per quintal, with a yield of 500 kg./ha. Some people cultivate sweet potato which yields only 1,000 kg./ha., sold for Rs. 50 per quintal. Others grow some horse gram (*kulti*). Here again we find confirmation of the progress of pulses in Orissa, which are increasingly becoming a cash crop in view of their high price. Apart from recently dug wells irrigating vegetables, one does not notice much change. Holdings are often around 1 ha. and that too of rather poor soil, yielding poor crops.

Taking a small road we proceed further to Pipilia where some farmers have benefited from government schemes. This is the home of the Gonds (tribals) who have been cultivating their land with the plough and bullocks for a long time.

Chaitanya Mahapatra shows us his kitchen garden (0.4 ha.) irrigated by a recently dug well, 30 feet deep (Rs. 1,500 government loan and Rs. 1,500 as a subsidy). Several crops follow one another on very carefully cultivated plots. Cowpea, vegetables and chilli are harvested in September. From October to January cabbage and cauliflower are grown. Some papaya and lime trees are growing fast and will bring in additional income. During May-June, the supply of water falls. There is just enough for nurseries of bean, transplanted in July when the rains fall. It is a pleasure to meet Chaitanya. Being illiterate does not prevent him from being clever and industrious. He used to live extremely frugally off some maize and pulses grown on a few elevated plots. Now he gets enough money out of his kitchen garden to live decently. One married son helps him. A second son, supported by his father, is at college.

Chaitanya's neighbour also cultivates vegetables on 0.4 ha. irrigated by his well which is fitted with a 5 HP diesel pump. On 700 sq. m., he grows 3,000 cabbages, sold for Rs. 7,500, with production costs around Rs. 1,000. This is followed by other vegetables. He owns 1 ha. of rainfed paddy, out of which, with good rains, he gathers 1,500 kg. or a little more. Not far away, at Denkikot, the weekly bazar is replete with vegetables, fruits, sweet potato and pulses supplied by the neighbouring villages.

Not all people could benefit from such schemes and it is also not certain that one could dig wells indefinitely for fear that water may not be replenished. Already, the water supply of all wells falls sharply during the hottest months of the year before the monsoon. Many people own some land, but also have to work as agricultural labourers. However, there is not much opportunity for employment as there is little double-cropping. Besides, they earn only Rs. 5 per day, along with some food.

In Sankarpur, a small village 6 km. north of Keonjhar, we come across a similar pattern of agriculture : traditional rainfed paddy with fluctuating yields and some ordinary wells irrigating small patches of vegetables. Lachman, who belongs to the Munda tribe, earns Rs. 2,000 from cultivating 0.2 ha. of brinjal, followed by other vegetables. He got a loan of Rs. 2,000, with a subsidy of Rs. 1,000. Before this he was forced to survive with only some maize and pulses.

In another village, we stop at the house of a relatively big landowner, Krishna Chandra, who owns 4 ha. of land. He lives with his wife, two young daughters and a small boy. On 3.2 ha. he grows paddy, the yield of which fluctuates between 1,000 and 1,500 kg./ha. With good rains and some chemical fertilisers, Krishna Chandra could get up to 1,750 kg./ha. 0.8 ha. Elevated land with very poor soil accounts for 0.8 ha. which remains fallow except when used for cultivating some pulses. There are no cows but he has six bullocks. Krishna Chandra has two permanent labourers who receive 400 kg. of paddy each, per year, and three meals a day. They are also given some clothes, and, if necessary, some medicine. These men enjoy a certain amount of security, but for extremely low wages; 400 kg. of paddy being equivalent to about Rs. 600, i.e., Rs. 50 per month and some food. Casual labourers receive 4 kg. of paddy for transplanting and 7 kg. for the harvest, in both cases without food. Krishna Chandra's family has no paddy left for sale after covering their own needs. They have a kitchen garden, not very well kept, so that they are able to sell vegetables. Their small pond supplies some fish for home consumption. The whole family is helped by the eldest son who earns Rs. 650 per month as a peon in Keonjhar and sends money home.

With 1 ha. of land, Gamanq Behera cannot maintain his family of six adults and five children. Fortunately one of his sons works as a peon at Keonjhar and saves Rs. 200 out of Rs. 500 every month.

Mahanta, like his two brothers, inherited 0.4 ha. from his father. He has developed a good kitchen garden with the help of his well. He also works as a casual labourer for Rs. 5 for transplanting and Rs. 7 to 8 with some food for harvesting. He has no cow, but three bullocks which enable him to take 0.8 ha. on a sharecropping basis. He bears all the costs and keeps half the paddy. The whole family of ten people manage reasonably well. Each of them have a few sets of clothes, they wear sandals and together they own a cycle.

After passing through rocky plateaus and areas which, having benefited from afforestation schemes, were abundant with sal, eucalyptus and cashew nut trees, we finally stop 20 km. off Keonjhar at Badamuli. Here, the cultivated area of 80 ha. is divided into 100 cultivators' households. Thirty-three families are landless. Such data give us an idea of the population pressure even in such areas. Yet, poverty is not too acute because no less than 100 men are working at the construction of the new port and industrial city of Paradeep for seven to eight months every year, earning Rs. 12 to 15 per day.

We see some kitchen gardens around wells (there is a weekly market 4 km. away), rainfed paddy fields, horse gram (*kulti*) on higher grounds, etc. The cropping pattern is not very different from the other villages. Daily wages amount to 3 kg. of paddy and some food for harvesting and a little less for transplanting. We come across Nayan Mohanta, the oldest man in the village who is eighty years old. 'What have been the changes here?' I ask. The village, he says, is no more isolated. The road and some buses connect it to bigger villages. A school has been constructed. Paddy yields have increased from 750 kg./ha. to 1,250 kg./ha. in good years. A few farmers now use small doses of chemical fertilisers. Then, many men have found employment in Paradeep. Two hand pumps have been installed to overcome the problem of drinking water in summer when the nearby tank is almost empty.

To summarise these observations, let us underline the following points. Road infrastructure has expanded considerably and this was already clear during my first visit in 1984. We found some main roads asphalted, small village roads, both of which contributed to the promotion of trade and bazars. Paddy is progressing very slowly. In many places yields have, for a long time, been merely fluctuating at around 1,000 to 1,500 kg./ha. and sometimes less. Plots yielding an average of 2,500 to 3,000 kg./ha. (using new seeds and chemical fertilisers) remain rare.

It seems that there has been some expansion in the area devoted to pulses. New wells irrigating kitchen gardens have a sizable impact on the living conditions of their owners, but unfortunately, there are not too many of them.

There is a substantial variation in wages :

	Daily wages *1975*	*1985*
Near Keonjhar	Rs. 2.00 – 3.00 *or* 3 kg.(paddy)	Rs. 5.00 – 8.00 *or* 4 – 8 kg. (paddy)
15 – 30 km. away	(data not available) 2 – 3 kg.(paddy)	Rs. 5.00 – 6.00 *or* 2.50 – 4 kg. (paddy)

Usually, near Keonjhar the wages do not include food, whereas further away, some food is often added. As for the wages in the mid-1970s, I cannot be too accurate. These data were obtained from the farmers when I asked them about the situation about a decade ago. If these data are correct there seems to have been a slight improvement in real terms. In some areas all wages are given in kind, in others all in cash, or in cash for all work excluding harvesting.

A medium irrigation project [7]

While ordinary wells and low-lift pumps can be used for irrigation, they are limited in scope so that, wherever possible, medium and major irrigation projects are needed. Several of them are under construction in the interiors of Orissa, in several cases with the assistance of the World Bank.

The river Kalinjar, a tributary of the Baitarani, is being dammed under one such medium irrigation project, 18 km. from Keonjhar:

Catchment area	358 sq.km.
Length of central spillway	118 m.
Length of earth dam	1,126 m.
Cultivable command area	9,800 ha.
Area irrigated in *kharif*	9,300 ha.
Area irrigated in *rabi*	5,900 ha.
Length of right canal	15.80 km.
Length of left canal	21.10 km.
Distributories and minors	110.45 km.

Population covered 31,850, of which 21,400 belong to Scheduled Tribes.

[7] A brief note on Khanjari irrigation project, Keonjhar, 1975.

The original cost (at 1979 prices) was estimated at Rs. 138 million. The work started in 1980 and the project was scheduled for completion in 1985. Various delays occurred, particularly with regard to land acquisition, so that the project is scheduled to be completed sometime in 1987. The final cost should amount to Rs. 250 million. The project includes the final water courses to the fields with turn out points up to plots of 5 ha., so that wasteful field to field irrigation can be avoided.

The necessity of such works cannot be questioned. After all that we have seen of poor rainfed paddy yields, irrigation will act as a protection in case the rains fail or are delayed and will enable a good crop in the *rabi* season.

Two basic issues remain : time and money. In many developing countries such a scheme would be considered quite large, whereas in India it is classified as a medium project. Now, if we assume that there will be no further delays, it will have taken at least eight years for the construction. Second, the cost is very high : roughly Rs. 25,000 as investment per ha., whereas in the Gangetic basin, as a result of private tubewells, the investment comes down to Rs. 3,000 to 5,000. The rocky terrain, undulating plateaus and hills involve expensive construction work unlike construction on alluvial and flat soils. The isolation of the area is another problem with regard to approach roads, electricity, residential and office buildings on the dam site. Finally, how secure will the project be? If a severe drought occurs, the reservoir will enable farmers to irrigate half the scheduled area during the *kharif* and there will be no water left for the *rabi*. Last but not the least comes the problem of proper levelling of the fields, an expensive task in this type of terrain.

The view from the district

The district,[8] after a long period of stagnation, is no longer dormant. But the awakening has come late, at a time when the population pressure is becoming heavy. The scope for agricultural development is limited by severe natural constraints. Soils are frequently porous which reduces their water retention. The profile

[8] Data supplied by the district authorities as well as, Keonjhar *District Statistical Handbook, 1978-79*, Keonjhar Statistical office, 1982.

of undulating plateaus encourages the run-off of water. The flow of the rivers is affected by erratic monsoons. Ground water is often deep and the cost of digging wells is high in such a terrain. Besides, the potential for irrigation (surface and ground water) is not much. For all these reasons, irrigation remains confined to a small area:

1983-84

Net cultivated area	287,000 ha.
Kharif	279,000 ha.
Rabi	84,000 ha.
Gross cultivated area	363,000 ha.
Irrigated area *kharif*	25,000 ha.
Irrigated area *rabi*	9,250 ha.

The total irrigated area was 20,000 ha. for *kharif* and 7,420 ha. for *rabi* in 1979-80.

Means of irrigation (ha.) 1983-84

	kharif	*rabi*
Major, medium irrigation	14,700	380
Minor irrigation	1,500	1,166
Low-lift pumps	1,800	1,095
Ordinary wells	2,000	1,440
Others	5,580	5,170

With such a poor rate of irrigation, it is no wonder that the consumption of chemical fertilisers remains extremely low.

Consumption of chemical fertilisers in tonnes of nutrients

	1979-80	1983-84
Kharif	608	984
Rabi	298	852

The consumption of pesticides too, is pratically nil. Rainfed crops remain extremely vulnerable to weather conditions. For the past eleven years, we have the following records:

Rainfall for Keonjhar district (mm.)
(Normal rainfall : 1,534 mm.)

1975	1,425	floods
1976	931	drought

1977	1,295	drought
1978	1,204	tornado
1979	1,150	drought
1980	1,259	drought
1981	1,277	tornado
1982	1,363	drought and cyclone
1983	1,787	drought and cyclone
1984	1,655	drought and floods
1985	1,891	floods

The table does not indicate the dispersal of rains, hence there is a possibility of drought even with 1,259 mm. (1980), when rains may have failed at a crucial period of paddy growth. We see also that cyclones or tornados can occur. As for floods, if they damage monsoon crops, they favour crops in the dry season thanks to increased soil moisture.

Rains affect both areas and yields. For instance, in the two years 1974 and 1975, paddy (around 66 per cent of the cropped area) covered 178,000 ha. in the monsoon, which fell to 143,000 ha. in 1976, due to a very severe drought. Yields vary a great deal from a district average of 800-900 kg./ha. in a bad year to 1,600 in a very good year. The bulk of paddy is cultivated during the monsoon (1,50,000-1,80,000 ha.) versus 30,000-40,000 ha. harvested at the beginning of the monsoon and 1,000-2,000 ha. in the dry season.

In view of these circumstances, could one not envisage a different crop pattern since paddy requires so much water? Even where feasible, such changes do take time, especially when they occur under rainfed conditions which in any case prevent high yields. So far, millets cover about 2,000 ha., and their prospects are limited since they are adapted to low rainfall. Maize (5,000 to 9,000 ha.) is sensitive both to excess water and drought, so that it is not certain that it can expand on a large scale. The situation seems more promising for a shift from paddy to pulses and oilseeds in areas which are not prone to floods. The former have increased from 8,000 to 23,500 ha., the latter from 2,200 to 11,300 ha. between 1978-79 and 1983-84, at the expense of land devoted to paddy. Vegetable production has progressed (9,500 ha. to 15,600 ha.) due to irrigation. Better soil management practices, proper utilisation of excess run-off water could help earlier maturing varieties of paddy. What is disturbing about all these tasks is their complexity and in some cases, their cost.

The division of holdings, considering the existing yields and their rather limited prospects, represents another matter of concern. Of the total agricultural holdings, 73 per cent were below 2 ha. and 42 per cent below 1 ha. in the 1970s. The difference in the number of landless labourers *vis-á-vis* cultivators, however, is not as high as in the plains: 65,000 versus 156,000 (1978).

Once again comes the crucial necessity of a greater diversification of the economy. Forests cover 2,500 sq.km. out of a total district area of 8,303 sq.km. Large tracts suffer from encroachements, illegal felling, over-grazing. Clashes also occur between the forest department which wants to protect forests and the tribals who still practice their ancient technique of shifting and burning cultivation. There is, however, the possibility of gradually overcoming these defects and difficulties. One comes across more afforestation schemes than in the past, either under the forest department or on private land. A joint venture between the government and a private firm has begun tea plantations on 800 ha., 150 ha. of which are already producing tea. Eucalyptus, sal and cashew trees are also progressing.

Keonjhar, like several other districts, has introduced a scheme to help the local population produce *tassar* (a kind of silk). This is part of the All India Tassar State Project. The project in Orissa is supported by the Swiss government. In 1984, I participated in a joint Indo-Swiss evaluation. Such a programme brings clear benefits to the tribal population as a substantial addition to their income, but here again one faces the complexity of the tasks involved : technical problems such as the introduction of new trees on which are fed the larvas, social problems such as those between Hindu money-lenders or traders and the tribals, economic problems relating to loans-cum-subsidies and marketing. It is also not simple to achieve coordination between the forest department and the Tassar State Project officers. Considering the large number of tribals concerned, such a scheme cannot be expected to reach a sizable part of the people very quickly, although the results have been encouraging.

To replace the system of shifting and burning cultivation with settled agriculture or horticulture represents another complex problem. In 1978 I had seen such schemes in Phulbani district which looked promising. In 1985, I learnt that they had failed.

As we have seen, some Scheduled Tribes can adjust rather

easily to innovations and more intensive methods of cultivation, but there are others who are more isolated in the mountains. Some of them, like the Paharias of Deoghar, are in danger of extinction.[9] Others suffer exploitation at the hands of the high caste Hindus who look down on them. Some officials do the same, regarding them as 'naked savages'. Such problems seem less acute in Keonjhar than in parts of Vidarbha in east Maharashtra, but they do exist and could become worse in many tribal areas.

Activities outside agriculture and forestry play a minor role. The urban population was and remains one of the lowest in any district : 1.60 per cent in 1951, 7 per cent (i.e., 67,000) in 1971 and in 1981. The district contains mineral resources like iron ore, manganese and chromite,[10] but several of them are not exploited. Large scale industries could not be developed since a big integrated steel plant and other industries have been set up in Rourkela, a neighbouring district. The lack of rail facilities should also be taken into account.

As in other districts and states, the IRDP and other schemes for the poor are in operation. The stories of their success enumerated to us during our field trip are encouraging. One should also mention that drinking water facilities have progressed thanks to a new model of the hand pump which can draw water from a depth of 20-30 metres, but which amounts to Rs. 12,000 to 20,000. A total of 3,840 units have so far been installed in 1,461 villages out of a total of 2,120.

In four years, the IRDP spent Rs. 21 million in subsidies and Rs. 34 million in credit for wells, milch animals, small tea plantations, silk, etc. As elsewhere, positive results are accompanied by various types of leakages and poor recovery of credit. Ownership rights on their homesteads have been given to 14,000 families between 1972 and 1984, usually on land that formerly belonged to the government.

Family planning is encouraged but its progress is slow. Because of this it is difficult to account for the pattern of population growth : 26 per cent and 28 per cent increase in the two decades

[9] See Chapter 11 of my book, *India's Changing Rural Scene* (New Delhi, Oxford University Press, 1982).

[10] Keonjhar belongs to this vast area so rich in mineral resources which includes parts of Bihar, West Bengal, Orissa and adjacent parts of Madhya Pradesh where many public sector industries were introduced in the late 1950s and 1960s.

1951-61 and 1961-71; and only 16 percent in 1971-81. It seems doubtful that family planning has had such a tremendous impact. Can one assume then, a relatively high rate of migration to more developing districts?[11] Or should one assume that the death rate has risen again?

During discussions with several district officers and visits to villages with them, I was on the whole impressed by their experience and ability. The district magistrate is an example of a young and dedicated officer. As in other districts, here too one feels more concern and commitment on the part of these officers than in earlier days. However, certain difficulties, some avoidable, others not, do arise. Law and order and natural calamities can suddenly absorb much of the time of the district authorities. Often the visits of ministers and other politicians are time consuming. Yet, I wonder whether the major constraint is not the size of the district. It is difficult to reach the remote areas and yet, it is necessary to visit them frequently as they require the closest supervision. If helicopters were cheaper, this would be the ideal means of communication! As it is, one spends so many hours driving in order to spend just a few hours on the spot.

Both districts have certain similarities. Physical constraints in terms of water and soils are serious. Irrigation can certainly progress further, but it cannot, even with massive investments, reach a higher percentage of the sown area. Paddy remains, by far, the dominant crop. With the paucity or uncertainty of water, yields are bound to rise only moderately. Because of this crops need to be further diversified. But once again, we cannot expect a miracle. On the other hand, the pressure of population on land is quite high, considering the existing low yields. The imbalance between agricultural production and population may further deteriorate.

All these facts make a greater diversification of the economy absolutely crucial. But here again, future progress does not look very smooth, regardless of the afforestation schemes or the secondary and tertiary activities accompanying a faster rate of urbanisation. As far as migration outside the districts is concerned, it is difficult to assess the potential. It will suffice here to say that in several other districts around Deoghar and Keonjhar, similar economic factors induce people to migrate to faster growing

[11] See the case quoted from one village of massive migrations to Paradeep.

areas. Over and above all these problems is the particularly difficult situation of the isolated *adivasi* tribes. Their situation rightly provokes growing concern not only on the spot, but as far as New Delhi. Finally, with reference to daily agricultural wages, we have noted some clear differences which depend on the proximity of these areas to the cities. It seems, but this would need confirmation through more detailed studies, that wages in real terms have not been declining. They may have, in certain areas, even increased slightly. However, the differences in wages between these and the more advanced areas, is very evident.

Chapter Eleven

In Semi-Arid Lands

A very large part of peninsular India falls under semi-arid tropical conditions with a rainfall ranging between 500 and 800 mm. a year, which prevents the cultivation of paddy except with irrigation. The dominant crops are drought resistant cereals, mostly *jowar* and *bajra*. In certain areas oilseeds, cotton and pulses are cultivated, whereas the few irrigated areas are devoted to paddy, sugarcane, vegetables and orchards of grape and other fruit.

A brief glimpse at Kurnool district

When driving from Hyderabad to Bangalore and Madras, we travel through typical semi-arid tracts on the eastern side of the Deccan in Andhra. On the outskirts of Hyderabad flourish small estates of vineyards surrounding pleasant bungalows covered with bougainvillaea. A few irrigated hectares bring a high return to their owners, who have been developing vineyards for the last twenty years. A similar trend is observed around Bangalore, in Maharashtra, and more recently near Delhi.

This fertile landscape does not last long. The horizon becomes flat with barren fields, red soils where *jowar* has been harvested and black soils where *jowar* has been sown at the end of the monsoon.[1] (Our journey took place at the beginning of December.) Trees, like villages are rare. Then, when approaching Kurnool, one sees wide and fine fields of tobacco.

In Kurnool,[2] the normal rainfall amounts to 620 mm., whereas the minimum requirement for normal cropping is 610 mm. The surplus is marginal and not infrequently, the rainfall is below normal, or the distribution is not uniform. The district is considerably

[1] *Jowar* is sown in light red soils with the advent of the monsoon. Heavy black soils are difficult to till at that time, so that farmers sow *jowar* towards the end of the monsoon and harvest it in the *rabi* season.

[2] Data from Kurnool *District Statistical Handbook, 1983-84*, Kurnool Statistical Office, and the district authorities.

large : 17,600 sq.km., which must make the task of the administration and the implementation of development schemes very difficult. With 2.4 million inhabitants (1981), the density per sq.km. is 136. The cropping pattern is typical of these areas:

1984-84	
Net sown area	880,000 ha.
Gross sown area	960,000 ha.
Net irrigated area	117,000 ha.
Gross irrigated area	158,000 ha.

Paddy, mostly irrigated, is confined to 100,000 ha. *Jowar* and *bajra* are subject to substantial variations following good or weak monsoons : 310,000 ha. in 1979-80 as against 265,000 ha. in 1983-84. Oilseeds amount, for the same years, to 172,000 ha. and 228,000 ha., and cotton to 112,800 ha. and 76,000 ha. Pulses cover around 160,000 ha.

Yields of main crops (kg./ha.) 1979-83

Kharif paddy	2,400 – 3,000
Rabi paddy	2,700 – 3,600
Kharif jowar	800 – 1,000
Rabi jowar	850 – 1,100
Kharif groundnut	500
Rabi groundnut	1,800
Cotton	100

As far as irrigation is concerned, the bulk of it relies on canals depending on the Tungabhadra and its tributaries. Major and medium works have reached the upper possible limit, whereas minor irrigation could expand to cover double the area, i.e., an increase of 30,000 ha., which is not much.

The constraints faced with regard to irrigation, particularly ground water irrigation, are even clearer in other districts of Rayalaseema. In a study on Chittor district, W.K.Olsen reached two conclusions. Over the period 1945-54 and 1975-84, 'there is no overall decline in annual rainfall.' In some areas, there has even been some increase during the last decade. 'Yet', the author adds, 'everyone from *coolies* to officials can see that there is a growing water crisis in Rayalaseema.' On the other hand, a vigorous expansion of borewells-cum-pumpsets has occurred, particularly in Chittoor between 1968 and 1984. But, 'wells are going dry, new bores are not reaching wet layers.' All this means an over use of water, so that recharge is not assured.[3]

While I was in Hyderabad, I visited ICRISAT (International Crop Research Institute for Semi-Arid Tropics), where research and experiments are conducted in dry land agriculture. The Institute breeds new varieties (preferably of short duration) of millets and pulses in particular. The aim is also to find methods of water conservation and soil protection, in order to check the high run-off of water. On even very mild slopes, the violent, though not frequent rains slash the soil and provoke erosion, which of course increases on steeper slopes. Agricultural watersheds are employed (3.10 ha. to a few hundred ha.) so that surface water is drained to a single outlet leading to a small tank where water is stored and can be used to complement failing rains. Contour ploughing along bed and broad furrow also contributes to check the run-off of water. For that purpose, a wheeled device with two ploughs drawn by bullocks has been designed.

On experimental fields at the ICRISAT, this package of new techniques, combined with interplanting of pulses and *jowar* or maize, produces good results. In the fields, as the district officers told me, the reproduction of this scheme is far from easy. Fairly large watershed development projects may cost, including soil conservation, Rs. 2,000 to 3,000 per ha. as initial investment, a large amount. More difficult is the necessity to achieve group action, since even on a watershed of 3-10 ha., several landowners will be involved. As to the new bullock drawn wheeled tool, it either requires a pair of high quality, very strong bullocks as used at ICRISAT, or two pairs of local bullocks, which is another constraint.

Various difficulties occur in other districts where the advantage of the new system does not necessarily bring enough increase in yields as compared to traditional methods.

New seeds and chemical fertilisers would affect yields positively provided the rains are good, a phenomenon that is not predictable. Hence, the reluctance of farmers to take risks.

For all these reasons, one gets the impression that a major breakthrough is not immediately in sight, a matter to which I will

[3] See W.K. Olsen, Manmade Drought in Rayalaseema, *Economic and Political Weekly*, 14 March, 1987.

return in chapters 13 and 14. Two trends are clearly evident in this area : the limited scope for much more irrigation and the complexity of rising production under rainfed conditions — a task much more difficult than what has been done in irrigated areas of advanced districts or what could be done in the eastern plains which enjoy such a high irrigation potential.

Proceeding further south, the landscape does not change much, with the exception of some irrigated areas (with tubewells and tanks) where farmers grow sugarcane, banana and coconut and, near Bangalore some vineyards.

The long drive through Karnataka and Maharashtra (January 1986)[4] takes one through endless, slightly undulating plateaus in the grip of a very severe drought. Fields have not been sown or have given a miserable harvest. Even fodder is lacking, so that poor people are forced to sell their bullocks for a measly Rs. 50 per head.

No less than 25,000 employment programmes have been started in Karnataka to give the poor an opportunity to earn some money, but they seem to proceed rather slowly. Crash programmes for drinking water are faring better, but the water table is going down and this may create new difficulties in the hot season.[5]

The situation hardly improves on reaching Sholapur in Maharashtra. We find the same barren plateaus bordered by flat greyish hills, except in the prosperous canal irrigated areas well known for sugarcane. Poor soils, lack of water, vast distances to cover in order to reach scattered villages, some of them lacking even drinking water. In spite of the importance of industries and of the number of cities which absorb some of the agricultural labour,[6] the battle for future development will be tough and costly.

Karwad village in Satara district

Satara belongs to the rainshadow area of districts starting from Nasik down to Kolhapur. In a fine study of the Bombay Deccan,

[4] From Bangalore I moved to Tamil Nadu and later returned to Karnataka.

[5] *The Hindu*, 8 December, 1985.

[6] In Kurnool district the urban population was 24.52 per cent (1981), cultivators 26.1 per cent and agricultural workers, 46.7 per cent of total workers. Note the very high percentage of agricultural workers who should be less busy than in more productive areas.

including the adjacent districts of present Karnataka, Sumit Guha emphasises the long term deterioration process which started in the 19th century. On the one hand the population increased, on the other, the eco-system was altered severely. 'Over many millennia the formation, erosion and deposition of soils in the Deccan led to a pattern in which steep gradients possessed almost no soils, the more moderate slopes and relatively level lands had deep retentive soils,'[7] a pattern very well represented in Karwad. Gradually, 'cultivation moved from the deep soils of the valleys to the thin soils of the slopes' (*ibid* : 121). As a result, the cultivated area did increase, but production moved up much less due to the marginal reclamation of land. At the same time, grazing ground was reduced which led to the cattle being poorly fed and slowed down the increase in cattle. Often, tillage operations became more careless than in the past. Foodgrains like *jowar* and *bajra* were particularly affected. Cultivation techniques did not change, nor were new seeds introduced, unlike what has been observed in western UP before Independence.

Cotton did expand considerably in certain areas, but with very low yields. In some districts, newly built canals led to the expansion of sugarcane. The number of wells increased along the pattern observed in Karwad. Groundnut was introduced in some districts like Satara. Under such conditions, severe droughts led to frequent famines or, at least acute scarcity. The crop failure in 1876-78 affected 63 per cent of the population in the Bombay Deccan. Lack of grain and rising prices (an eight-fold increase in Sholapur district) led to a lot of suffering in spite of the relief work organised by the government (*ibid.*:185). The assumption by Blyn that from the early 1930s onwards, agricultural output per capita began to decline, seems plausible.[8]

This imbalance between resources and population was partly mitigated by off-farm earning with people migrating to Bombay or other cities, a point of growing importance with the advent of the 20th century.

We are now back to a more detailed survey based on four visits: 1964, 1975, 1979 and 1986. Karwad village is located on a metal

[7] S. Guha, *The Agrarian Economy of the Bombay Deccan,* (Delhi, Oxford University Press 1985) p. 121

[8] G. Blyn, *Agricultural Trends in India 1891-1947,* (Philadelphia, University of Pennsylvania Press, 1966), p. 120.

road (which was asphalted), a few kilometres away from the small market town of Koregaon, 30 kms. from Satara. The landscape is typical of these vast stretches of plains, eroded plateaus and flat rocky hills where the rivers are dry for a large part of the year and the crops are generally poor (*jowar, bajra*) except for the small patches irrigated by wells : sugarcane, orchards of banana and papaya, vegetables and vineyards. This situation is now changing with the arrival of the Dhom canal[9] which has led to a sharp increase in irrigation and a much faster process of growth than had been experienced from the 1960s till the early 1980s.

Of a total area of 672 ha., 606 ha. are classified as cultivable, out of which 92 ha. were irrigated in 1964. Red soils of low fertility account for 232 ha., 97 ha. are extremely poor, on slopes furrowed by erosion or on gently sloping plateaus which often have only a thin cover of grass. The rest of the land is made up of black soils in a small valley along a stream which is dry except during the rains.

The land-man ratio follows the classical pattern observed in other villages : growing population with increasing division of holdings. Yet, as in several other peninsular areas — though not in Kurnool — there are relatively few landless labourers:

Population

1951	907	1971	1,336
1961	1,173	1981	1,802

These figures involve rather high in and out migrations. A semi-nomadic group (Nandiwala, a Telugu speaking group) settled in Karwad between 1951 and 1961. They worked as artisans and petty traders. Later, they moved away but have now returned (seventy people). Recently fifty people in service in Koregaon have settled in Karwad and commute from there. Another 150 people had to leave their homes because of the Dhom irrigation project and have been resettled in Karwad. For all these reasons, the natural increase in population is much less than it appears in the table. In fact, family planning, already fairly advanced in 1964, has been

[9] It comes from a reservoir built on the Krishna near Wai. As seen in Karwad, the canal network is also affecting the wells positively, by enriching groundwater through the seepage of water from minor canals and watercourses. The main canal is lined.

progressing, helped by the active propaganda introduced in the district in the early 1960s. In 1986, I was told that the number of pupils in the first year of primary school had dropped from forty to twenty.

The number of agricultural holdings has risen from nearly 200 in the early 1960s to over 300 twenty years later, two-thirds of them below 1 ha., and only seventy above 2 ha. Landless agricultural labourers amount to thirty families or so, as against twenty-one in 1964.

The Marathas, the dominant caste, are well known for their hard work. For many years they have been struggling against difficult conditions created by nature. The Mahars (Scheduled Caste) have little or no land but, as will be seen, they have also been quite clever in improving their lot. The few Brahmins, fairly large landowners, left the village in the late 1960s. There are about eighty Muslims.

I could not estimate the extent of tenancy, but it is not of much importance. Besides, as can be observed in other places, it works both ways ; a large landowner leasing out some land to a tenant, or a small landowner getting a job outside and leasing out his land to a bigger landowner.

The barrier created by the Western Ghats stops much of the monsoon, so that Satara falls in the rainshadow belt. Koregaon *taluk* receives around 600 mm. of rain, mostly during the monsoon. The situation prevailing in the 1960s was as follows: *jowar* and *bajra* were by far the main crops. *Jowar* was grown in the *kharif* and part of it in black soils at the end of the rains. *Bajra* was grown entirely in the *kharif*. Yields had been oscillating for decades: 330 to 460 kg./ha. for *jowar*, rarely more than 270 kg./ha. for *bajra*. This was not enough to feed the village as a result of which they had to buy grain with the proceeds of other crops. The area devoted to groundnut expanded in the next ten years. All these crops were exclusively rainfed.

There were forty-four wells, the construction of which involved heavy expenditure. One metre deep begin layers of rocks, so that it takes twenty-five men and at least six weeks to dig a well 7 metres in diameter and 10 metres deep. During the past 10 years, the *mot* system (bucket pulled by bullocks) has been replaced by diesel pumps on ten wells. (In 1964, Karwad had not yet received electricity.) Some small dams (*bandhara*) retained the water from the streams during the rainy season.

On their irrigated plots farmers grew sugarcane, chilli, vegetables and potato, which fetched a good price. At the time of our first visit, small patches of vineyards had been planted, bringing a return as high as Rs. 12,000 per ha. Farm manure and chemical fertilisers were used carefully. The expansion of cash crops, even on plots of land below 1 ha., enabled a number of farmers to live reasonably well, with enough cash to buy grain.

Even in those days, out of 200 families, 125 men had found a job outside Karwad, many of them in Bombay and Pune which was becoming the second major industrial city of Maharashtra. Others worked in Satara or were in the army[10] if not the police. The overall standard of living was not much lower than Mirpur during the same period, in spite of the poor output of foodgrains. This was compensated for by cash crops produced on irrigated plots and the incomes of those who worked outside.

The advent of the canal

When revisiting Karwad in November 1975 and January 1979, I did not find substantial changes in the economy. After suffering severe droughts in the early 1970s, the village began to recover. Foodgrains remained at the same level as before. With such uncertain rains, farmers quite rightly did not take the risk of spending money on hybrid seeds of *jowar* and *bajra*[11] supported by chemical fertilisers.

Irrigation, on the other hand, had increased from 92 ha. to 132 ha. and had improved in quality. By 1979 all wells were equipped with pumpsets, twenty-four of them worked on electricity, which by then had reached Karwad and fifteen relied on diesel. There was no more potential left untapped and already the ability to replenish ground water was questionable. In the hottest months, April to June, some wells hardly worked at all.

In 1978-79, part of the irrigated land had been devoted to foodgrain : 28 ha. of wheat giving 1,500 kg./ha., a rather low yield which could fall if the weather turned too warm. (Winter is less

[10] Maratha Regiment. As for the Mahars, it is the only Scheduled Caste which has its own regiment, the Mahar Regiment.

[11] Hybrid seeds, unlike new varieties of wheat and rice must be replaced every year.

cold here than in the upper and middle Ganges basin.) Hybrid *jowar* covered 24 ha., yielding 1,750-2,000 kg./ha. with fairly adequate irrigation and small doses of chemical fertilisers. Paddy (8 ha.) gave a mediocre yield of 1,250-1,500 kg./ha.

Other irrigated crops had expanded somewhat : sugarcane, grape and more recently, papaya trees, chilli and vegetables. Potato on the other hand, had fallen from 31 ha. to 4 ha. following a pest attack. Each crop of vegetables gave the farmer a net return of Rs. 1,000 to 2,000 per 0.4 ha., papaya Rs. 10,000 and grape Rs. 5,000-6,000.

Some farmers had bought crossed Jersey cows. The main landowner had bought a tractor. Two electric flour mills had appeared along with some more retail shops selling groceries, cloth, batteries, spices, tea, etc. The number of men working outside had risen from 125 in 1964 to 170 in 1978.

Though substantial, the process of growth and diversification of the economy was slowed down by the stagnation of *jowar* and *bajra*, mostly rainfed, which explains — at least to me — why wages were not rising in real terms — from Rs. 2 in 1964 (in Mirpur it was only Re. 1), to only Rs. 4 in 1978-79 (without food) as opposed to Rs. 5 to 6 in Mirpur.

In 1986, the whole scene had changed considerably. The Dhom canal (major irrigation project) has reached Karwad, cutting through the very poor soil of the hill above the village.

	Area irrigated (ha.)
1964	92
1979	132*
1985	206

*The figure, given to me for 1979, seems too high when compared with the data for 1985.

	Means of irrigation (1986)
Wells	41 ha.
Bandhara	43 ha.
Canal	122 ha.

When the whole network of field channels is completed, another 50 ha. will he added. The canals have another advantage: the seepage of water contributes towards increasing the amount of ground water so that the supply of wells has improved.

As for agricultural equipment, there is still only one tractor, but one threshing machine has been introduced, as well as fifteen sprayers and three dusters for pest control.

Sugarcane has been the main beneficiary of the new irrigation projects : 60 ha. yielding an excellent crop of 100 t./ha. of cane followed by a ratoon crop at 75 t.[12] Wheat still yields 1,500-1,750 kg./ha. as in the late 1970s or may even be little less. Some groundnut fields are now irrigated (750-1,000 kg./ha.), onion and vegetables bring a net amount of Rs. 3,000 per 0.4 ha. (5,000 gross). Potato has reappeared on 25 ha., preceded by ginger (10 ha.). Hybrid *jowar* reaches 2,500 kg./ha. when adequately irrigated. An improved, non-hybrid *rabi jowar* has been introduced, yielding 700 kg./ha., rainfed, and 1,500 with light irrigation. Local, exclusively rainfed *jowar* remains around 300-500 kg./ha. The movement of manpower has become a two way process. When compared to 1979, the number of men working outside does not seem to have increased much, but now, for cutting cane the mills[13] buying the cane send in teams of workers who operate on a contract basis. They come with their wives and children, often as far as Vidarbha.

Trade is expanding. The major shop owner has got a motor-rickshaw to bring goods from Koregaon. One TV set has been installed in the school and one by a farmer. Some people are now shifting from the bicycle to the motor-cycle.

This overall expansion has had an impact on wages, which have increased in real terms between 1979 and 1986, after having remained stagnant till 1979.

	Daily agricultural wages (Rs.)
1964	2.00
1979	4.00
1986	10.00 – 12.00

Daily wages have increased in real terms, compared to the cost of living: the free market retail price of *jowar* was 0.94 Rs./kg. In 1985 it was 2.30 Rs./kg. The 1985 prices would have been lower had there not been a severe drought.

[12] Here there is more intensive cultivation and a better climate than in the Ganges basin.

[13] Local cane crushing and the manufacture of *gur* (raw sugar) in the households has entirely disappeared.

The way they live

Let us start with the poor. Several Mahars (Scheduled Castes) have their houses in a part of the village which, on the whole, is rather compact. In 1964, Govin Pandurang Sawan was living with his wife and his four young children. Most of his land (0.8 ha.), was devoted to *jowar* and groundnut. He was able to get only 400 kg. of grain, i.e., one-third of his requirements. To make both ends meet, Govin used to work for eight months with the main landowner, earning Rs. 2 per day. Following his employer's example (see below), he had planted 0.1 ha. with grape on an irrigated plot depending on his well for which he had got a cooperative loan. He also had one bullock and one heifer. At that time I wrote of him : 'Govin has not quite extracted himself from his state of precarious existence, but he is close to doing so.'[14]

Whenever I visited Karwad, I could see Govin gradually improving his lot. In 1986, we once again talked in front of his house. Though he is aging, Govin remains alert. His brother and he have installed a pump on their open well. He had just cut his old vineyards in order to replace them. Next to them onion, garlic and vegetables are cultivated following a crop of hybrid *jowar* grown in the *kharif* season. Govin has a cow, a bullock and a few chickens. His brother who lives next door, works as a tailor, earning Rs. 100-125 per month. In addition, he earns from some agricultural work. From time to time Govin is employed as a casual labourer at Rs. 10 per day, while his wife receives Rs. 6. They have improved and enlarged their house. They have several beds, metal vessels, several sets of clothes and under the verandah stands their bicycle. The three sons are now adults and work in a factory in Satara.

Sambhuji Malu Bhosle, a Maratha, shares a well-cum-pumpset with another farmer. Of his 1.2 ha., half is irrigated by the well. He grows vegetables, potato and *jowar* in two crops per year. The rest of the land gave a poor crop of *jowar* until the advent of the canal, which has enabled Sambhuji to more than treble his production. He now lives quite well and has managed to buy two crossed Holstein Frison cows with a cooperative loan-cum-subsidy. He gets a yield of 5 l. of milk per day per animal, part of which is sold.

[14] See Gilbert Etienne, *Studies in Indian Agriculture* (Berkeley, University of California Press, 1968), p. 248.

Several other farmers manage quite well on very small holdings. Shankar Narain Garge, also a Maratha, intensely cultivates his 0.5 ha. In 1971 he borrowed Rs. 4,000 from the district cooperative bank to install a pumpset on his well, so that he could push sugarcane, papaya and lime trees, garlic and other vegetables on 0.3 ha. On 0.2 ha. he harvests 6-7 q. of wheat following *jowar*. He borrows a second bullock from a friend on an exchange basis to plough his fields. His buffalo supplies milk. From time to time they are able to eat one of their chickens and some eggs.

These activities alone do not allow Shankar to maintain his old parents, his wife and their four young children. That is why he is involved in petty business, buying vegetables and selling them at Koregaon where he goes by bus with his baskets. This way, he manages to earn Rs. 100 per month.

In spite of the progress in irrigation, many people still have a large part of their land outside the command of the canals or of the wells. Sampat Dadu Bhosle owns 1.6 ha. of very poor land which yields hardly any crop. The remaining 1.4 ha. is better located, part of it near his well which was fitted with a pump in the late 1960s. When we met in 1979, he proudly showed us his papaya trees. He has had to cut them following a pest attack and has replaced them with two crops a year: potato and bean followed by maize, fodder and wheat. On 1.2 ha. grow *jowar* and groundnut, benefiting from the canal. Two light waterings and some chemical fertilisers result in 1,200-1,500 kg./ha. of *jowar*.

Sampat's daughter is married in another village. One of his sons works as a cleaner for a truck driver. Another works as a truck driver. Sampat himself has been working as a carpenter and mechanic in the village for the past ten years earning Rs. 2,000 per year.

Several other farmers with 0.5 – 1 ha. have been able to improve their living conditions provided they have a small irrigated plot and at least one family member has got a job outside, or some additional non-agricultural work in Karwad itself.

In 1964, Ramachandra Ganapati Bhosle's family (Marathas) was both an example of remarkable success and of a family enjoying considerable power in local affairs. The father was the chairman of the village council. One of his sons was the manager of the local cooperative society. Ramachandra himself started with 0.4 ha. of land to finally become, through hard work and initiative, the largest landowner with 15 ha. Though 60 years old, he closely supervised his fields. Nearly 5 ha. were irrigated by a well-cum-pumpset : 1.6 ha. of sugarcane, 2 ha. of grape and chilli, vegetables and banana trees on the other plots. His dry land unavoidably yielded poor crops of *jowar* and some pulses. Ramachandra had thirty agricultural workers under him. He was the owner of the only flour mill in the village and used to go to Satara to sell his grapes and vegetables in his old Ford, a pre-war vintage.

In 1979, Ramachandra was no more. The six sons had each inherited 3.66 ha., and each irrigated only half the land. One of them, a barrister in Satara, has his land attended to by a tenant. The others live in Karwad.

Seven years later, I met Gajanan, one of the sons. He must have sold some land as he now has 2.8 ha., now entirely irrigated by wells and the canal. Most of the land is devoted to sugarcane, the rest yields vegetables. Out of 0.2 ha. he grows 25 q. of onion sold at Rs. 80 to Rs. 100 per quintal, while his production cost is Rs. 1,000. Gajanan still enjoys a comfortable life, but neither he nor his brothers carry the same weight as their father, a typical case of how power relations can change.

One of the permanent labourers under Gajanan receives Rs. 12 per day without food, a job he needs badly because his own 0.5 ha. only yield poor *jowar,* not yet under the command area of the canal.

In less fortunate areas

Even in the 1960s, Karwad was relatively favoured by nature. It had some irrigated land and enough drinking water. At that time I visited other places where poverty was more pronounced. For instance, in Tadwale, Koregaon *taluk,* a population of 2,500 was supported by 800 ha., out of which only 65 ha. were irrigated. There were only 190 men working outside. In Phaltan *taluk,* with a rainfall lower than in Koregaon, the situation was so bad that it had been permanently classified as a 'scarcity area'. The soils were exceptionally poor. Out of 2,000 cultivated hectares, only 180 ha. were irrigated and that too not completely because wells were often dry. Besides, drinking water was very bad, so that a distribution system with pipes and taps was being installed.

In 1986, I made a long journey of 70 km., from Koregaon to Mhaswad through one of the most desolate landscapes I have ever seen in the Deccan. Most of the land is mildly undulating and barren : no vegetation, reddish grey soils, here and there a dusty track leading to a remote village. This is one of the worst drought-prone belts of Maharashtra where the Sir Dorabji Tata Trust began a rural development project as early as 1951.

The area concerned covers thirteen villages, 140 sq.km., 8,047 people (1971 Census) and a density of 57 per sq.km. Soils are extremely poor. The rainfall is around 300-400 mm., falling mostly in the later period of the monsoon, so that farmers cultivate *rabi jowar* and later, on some plots, groundnut. Yields of *jowar* can be as low as 150 kg./ha. As for drinking water, it is hardly available in summer. Under such circumstances, as many people as possible are trying to find a job in less poor regions.

The Tata project was later broadened with government schemes, especially the IRDP. The project acted as a catalyst and is still effective under its competent leader. He lives in the area, is sharp and has a deep concern for the poor. The area is now connected to the rest of the district by several metal roads and reasonably good tracks where five to six buses ply every day.

Soil improvement through contour bunding started in 1953 with the help of the government. Later, in the 1970s, several percolation tanks were constructed. Small dams were built on the streams to collect the rare water. The supply of drinking water began to spread. Electricity came later and stimulated the use of pumps.

In the 1970s, the benefits of the various programmes were emerging : the area devoted to *bajra* and *jowar* had been reduced in favour of better crops : wheat, sugarcane and cotton, with the help of irrigation.

The old dispensary has now been expanded. Housing schemes have started. Trade is expanding. The wages of casual labourers, practically non-existent till now for lack of work, amount to Rs. 6-7 (1985-86), which is not so bad for such remote places.

Let us stop at village Gangoti were eighty people live on 66.8 ha. Now 52 ha. are irrigated by a tank equipped with lift pumps. Water remains scarce so that light waterings have to suffice. Yet, the progress is considerable compared to the miserable yields of the past. *Rabi jowar* yields 1,000 kg./ha. It is followed (February to

June) by groundnut at 700-800 kg./ha., unshelled. Landowners with 2 ha. to 6 ha. can therefore manage fairly well. Animal husbandry is improving. Milk collection and sales have started.

In Parwan live seventeen families of Mahars, who are also emerging from their extreme poverty. A tank equipped with low-lift pumps and pipes (needed in these undulating plateaus) has brought under irrigation 12 ha. of land. With the assistance of the IRDP and supervised by the person in charge of the Tata project, the land has been levelled and bullocks and iron ploughs, spray pumps, chemical fertilisers and seeds are now available. The villagers supply the labour force for the infrastructure.

The benefits accruing from such activities cannot be doubted, but they raise two major questions: those of money and local leadership. The project in Parwan has, over five years cost Rs. 170,000, i.e., 10,000 per family supplied entirely as grants. The project in Gangoti amounts to a subsidy of Rs. 122,000 and a government loan of Rs. 228,000.

The total lift irrigation scheme in all villages can be summarised as follows:

Total bank loan	Rs. 580,000
Total government subsidy	Rs. 184,000
Total area irrigated	86 ha.
Families benefited	105
Bank loan already repaid	Rs. 221,000

Not included in this is the cost of manpower, given voluntarily by the villagers.

The cash investment per ha. amounts to Rs. 8,900. If we take only the subsidy it amounts to Rs. 2,140. In looking at the total expenditure over twenty-seven years it amounts to about Rs. 10 million, i.e., Rs. 1,240 per head (1971 Census), an amount supplied by the Tata Trust and increasingly by the government. The total investment has provided employment for 300 to 400 workers for eight to nine months.

No less important is the calibre of the people responsible for the project. We have already mentioned the leader of the Tata project who was also functional in stimulating the district administration. Obviously it would be difficult to reproduce on a massive scale the same concentration of talent and money.

Satara district

Here is another big district covering 10,548 sq. km. with a population of 2.04 million inhabitants and a density of 190 per sq. km. Due to sloping lands and erosion, a large part of the land is too poor to be cultivated.

Total cultivated land (net) 575,000-650,000 ha.

The rainfed area varies following the monsoon:

Kharif jowar	80,000–86,000 ha.
Rabi jowar	75,000–101,000 ha.
Bajra	82,000–100,000 ha.
Groundnut	51,000–60,000 ha.

Soil conservation and improvement is crucial in view of the limited possibilities of irrigation. In the 1960s contour bunding was considered the major remedy against erosion and the run-off of water. Several schemes were started with government assistance. That method has proved inadequate and is being replaced by more efficient techniques such as the one devised by ICRISAT. Contour ploughing is being emphasised more than ever. Such changes are rather slow to come about due to the complexity of problems involved.

The area of irrigation has also been rising:

Irrigated area in ha. (net)	
1960-61	71,000
1970-71	80,400
1978-79	98,300
1984-85	138,000

The recent acceleration is, to a large extent, due to the Dhom canal, which enjoys a potential not yet fully utilised, of 30, 400 ha. in Satara district. A lot of land still needs levelling, a costly task in such terrain that is often hilly and rocky, unlike the alluvial plains. Wells-cum-pumpsets account for 45,000 ha. Total irrigated land accounts for 17 per cent of the net sown area. The upper limit can be raised to 27 per cent but investments will be heavy. A well-cum-pumpset amounts to Rs. 25,000-30,000 with a command area, from July to March, of 2 ha. and, 0.5 ha. in the following summer months. Drought in two consecutive years causes serious trouble as the water table falls dangerously.

Irrigation has benefited many crops. Irrigated sugarcane has increased to 28,300 ha. in 1985-86 from 17,000 in 1978-79. Irrigated *rabi jowar* (often hybrid) accounts for 40,000 ha. now as against 27,500 in 1978-79. Paddy has increased from 28,400 ha. to 35,500 ha., and more than half is now being irrigated. Wheat has now reached 15,000 ha. Potato and vegetables have also increased. Sunflower is cultivated on 2,600 ha. Papaya and mango trees and pomegranates have made some headway.

The progress of irrigation goes along with a greater use of chemical fertilisers:

Consumption of chemical fertilisers in tonnes of nutrients

1963-64	1977-78	1983-84
900	8.260	30.856

A new variety of groundnut has been released, the yield of which has risen from 1,100 to 1,800 kg./ha., unshelled. It can be followed by *rabi jowar* on the same soil. While rainfed crops remain at a low level, lightly irrigated *jowar* yields around 1,500 kg./ha. Sugarcane, as seen in Karwad, is doing well.

Foodgrain production is often erratic because of the predominance of rainfed crops:

Production of foodgrains in tonnes

1960-61	296,000
1977-78	384,000
1983-84	417,000

All these years enjoyed excellent weather conditions, but in between there have been several setbacks. For instance, in 1978-79 production fell by 50,000 t. The population has increased by about 50 per cent as against an increase of 41 per cent in foodgrain. In the three years, 1984-85 to 1986-87, production has been below the peak of 1983-84, due to severe drought.

Wherever irrigation is relatively important as in Karwad, we have seen that many farmers compensate for their deficit in foodgrain by the sale of cash crops, something that is not as feasible in the many villages with less irrigation and a lower rainfall. The western part of Satara receives (in a normal year) 650 mm of rain, whereas the eastern part gets 450 mm.

In 1979 itself dairying was quite progressive, a trend which continues today. In 1984-85, 200,000 l. were collected every day.

Cross-bred cows, as seen in Karwad (Jersey, Holstein-Frison), are less rare than before, providing several litres per day unlike the local cows which give 1 litre.

The progress of vineyards and other fruit is hampered by various difficulties. Papaya trees and vineyards are exposed to pest attacks. In fact, vineyards should be given up in Koregaon in favour of drier places. On the other hand the expansion of sugarcane is encouraging.

Vegetables are exposed to great price variations, due to either a glut or a shortage in the market. They are also vulnerable to disease. Good commercial networks and quick delivery systems are needed, something relatively easy in Karwad, but much less so in remote villages. All these practical difficulties show that even this small but crucial sector is not free from pitfalls.

Several afforestation schemes have been started by the government or are being encouraged through loans.

What about job opportunities outside agriculture? The town of Satara keeps on expanding with new housing settlements, a growing bazar etc. Sixty new medium industries have been created in the district during the past few years with 50 to 100 workers per unit. There are now seven sugar mills as against three in 1978. Koregaon too is becoming a small town with some modern shops, small industries and mechanical workshops. Although the rate of urbanisation remains rather low (15 per cent), out of an active population of 684,200 (1981), 29 per cent are working outside agriculture. We must not forget the rather large number of people who keep going to Bombay (300 km.away from Satara) or to Pune (100 km.), which has become a large industrial city with 1.7 million inhabitants (1981). Nasik, further north, is also beginning to attract industries. So also is Thane, on the outskirts of Bombay.

On the whole, Satara is well located, near the main industrial belt of Maharashtra, enjoying a number of outlets for labour which becomes more limited when we move further away from the triangle — Bombay – Nasik – Pune. One should also not forget the army and the police who recruit many Marathas and Mahars.

As far as daily wages in agriculture are concerned, the examples given in the field fit rather well within the overall picture in the district : Rs. 10-12 where irrigation is fairly advanced, Rs. 5-7 in less developed areas. Besides, the more area covered by irrigation, the more is the employment created.

The administration makes a good impression on the whole. The system of panchayati raj which had been given so much scope and influence in so many states in the 1960s is now dormant. This, in my opinion, is not a tragedy, because even in its heydays, most of the actual and useful work was done by the administration. In that regard, the system adopted in Maharashtra differs from other states, except Gujarat. While the district magistrate or collector is the head of the district, another IAS officer of equal rank, the chief executive officer (CEO) is in charge of development in the district. In 1964, the CEO was particularly outstanding, but immediate subordinates rather mediocre. Today, the whole staff seems more qualified. The Training and Visit System functions rather well. The IRDP officer is dedicated to his work and knows his job. Although political leadership in Maharashtra has considerably deteriorated when compared to the period up to the end of the 1960s,[15] the administration nevertheless remains at a rather high level and compensates, at least in part, for the political instability.

[15] The situation seems to be improving with the advent of S.B. Chavan as Chief Minister in 1986.

Chapter Twelve

Deserts, Lack of Water and Poor Crops

Already precarious in semi-arid zones, the balance between men, animals and natural resources becomes even more fragile when we enter the arid, semi-desert or desert regions of western India.

A glimpse at Kuchch

Starting from Bombay in the middle of January we drove to Ahmedabad and then to Bhuj in Kuchch, near the Pakistan border. The contrasts in Gujarat's agriculture are striking. Kera district is famous for its prosperous farmers — Patidars. With its dairy schemes, sugarcane and banana plantations, tobacco and cotton, this district is thriving. In Mehsana district, the tubewells have multiplied since my first visit in 1957, allowing vast fields of wheat to be irrigated. Lively bazars and small industries are other signs of activity. Roads, both state and district roads, are of a standard rarely reached elsewhere in India.

The other side of the coin, however, should not be overlooked either : poor eroded soils, barren land, empty fields where nothing can be sown because of the drought. Semi-desert conditions prevail the nearer we got to Bhuj, from where we proceeded to Khawda. Herds of camels can be seen on the road, eating small shrubs. When it rains, villagers sow some *bajra*, which they could not do even in 1985. They rely more on animal husbandry. Very marginal rainfall can provide enough fodder so that a cow can yield around 5 l. of milk per day, which falls to 1 or 2 l. in case of drought. In Khawda, the dairy society is expanding. Several people make *ghee* (clarified butter) for sale. Others earn money by grazing cattle which are sent from Saurashtra for a few months in the year.

Handicrafts are more alive today than they were ten years

ago.[1] They dye and print cloth with wooden blocks, embroider cloth fitted with small mirrors, make ladies hand bags, *razai's* (thin quilts blankets stuffed with raw cotton) leather shoes and pottery. Trade brings some income[2] but so does smuggling to Pakistan. All in all, one does not come across acute poverty : traditional circular houses with circular thatched roofs are well kept with some furniture and fine granaries for storing grain.

Khawda enjoys a drinking water scheme, a deep well, a water tower and taps to which women come in the morning and evening to fetch water. They are vividly dressed in flared dark red skirts, their shining brass vessels on their heads. And then follows a lively gossip session while they wait to collect water.

The landscape is dominated by small hills where shrubs grow between the rocks. We walked up to Payia (5 km.), encountering camels loaded with wood or grain on the way. Drought conditions were severe all over, but several relief works had been opened. Most of the people of Payia, including women, were building a reservoir to collect the rains. They earn Rs. 7.50 to Rs. 11. I did not hear any complaints about malpractices.

One day we went north towards the Rann of Kuchch. We rode superb camels, past remote villages, until we came upon the most amazing sight : nothing but a white, almost snow-like flat plain, which is in fact a crust of salt. Before the salt desert we stopped at Dinara where 3,600 inhabitants live in several hamlets. If rains are adequate they sow 2,000 ha. of *bajra* which yields a few quintals per ha., but 1985 was the second year that went by without rain.

This brief visit to Kuchch was impressive. This is one of the most isolated places in India where one does not come across real misery, but rather, people who are happy, clever and hospitable. Most of the inhabitants are Muslims (Mohman)[3] who do not practice the rigid separation of sexes (*pardah*) as found in the more advanced societies.[4]

[1] Micheline Centlivres, a Swiss anthropologist who visited Khawda in 1976 found only one household involved in printing cloth. There are a few more now.

[2] Before partition, the area had close contacts with Sind. The opening of a new crossing point at the Indo-Pakistan border will further help trade.

[3] Mohman is one of the quasi castes of Islamic society. From Kuchch and other parts of Gujarat have come some of the people who, from trade moved into larger business and industry in Pakistan. Some of the biggest industrialists in Karachi are Mohman.

[4] My daughter, Danielle, brought us here after having visited Kuchch several times herself.

In Bhuj, I gathered some general information about the district. The district covers 45,652 sq. km. (almost the size of Switzerland) with a population of one million in 1981. Around Bhuj grow some irrigated wheat and cotton. Occasionally *bajra* and *pulses* are also grown. Animal husbandry (1.4 million cattle, goats, sheep, camels) plays an important role in the economy.

How can one run such a huge district? A young IAS officer is facing this formidable challenge quite well. The government is increasing development schemes. Several roads have been built (partly for strategic reasons in this border area). The drinking water supply, as seen in Khawda, is improving. As a result of two consecutive years of drought, several relief projects are now in operation, so that in Khawda and around one does not come across people starving.

Around Jodhpur

The road fron Ahmedabad cuts through fertile land, with irrigated fields of vegetables and wheat. Men earn Rs.14 per day for collecting vegetables, a high wage typical of areas near a large city. For the wheat harvest they receive Rs.10 without food.

Further north one encounters hilly terrain. Here tubewells or wells irrigate the small valleys where wheat and mustard grow together. Near Udaipur, we even saw the old Persian wheel described by Babar in his memoirs. The equipment is made of wood, water is collected in earthen pots tied with ropes on the wheel.

When we cross the Aravalli range we enter the driest part of Rajasthan. There are a few scattered pockets of irrigated wheat, otherwise, between Pali and Jodhpur, the landscape is barren. Several of the roadside villages have a water tower and a pipes-cum-taps system.

When I first passed through Jodhpur in 1959, it was not exactly a dormant district, but it was awakening only slowlv. Roads were rare and the few that were there were in poor condition. Irrigation pumps were practically unknown. During my first survey in 1979 many changes had occurred. However, in view of the enormous variation in crop yields from year to year, the overall economic situation is susceptible to considerable change. In 1979, I was in Jodhpur after a good monsoon, unlike in 1986.

As far back as 1979, the degree of diversification of the district economy was striking, enabling even poor people to avoid acute poverty in spite of so many natural constraints: semi-desert conditions with rugged hills, wide open uncultivated tracts, sometimes broken by sand dunes, grazing ground for cattle, sheep and goats when the rains come (normal rainfall 320 mm.). *Bajra* could yield as much as 500 kg./ha., but more often it would not go beyond 200-300 kg./ha. or even less. And, of course, in case of severe drought, farmers would not even sow grain, or would lose it entirely if the rains failed after the first showers. Tubewells were in use in several areas and allowed a relatively good yield of crops if the water could be replenished. Green patches of new seeds of wheat, chilli and citrus and mango trees mingled with the landscape dominated by light brown to greyish hues.

The dairy industry and milk collection played an increasing role with a wide network of trucks going into villages and then bringing the milk as far as Delhi.

Handicrafts, some small industries, workshops for the numerous tractors in use, the progress of roads, tourism, stone quarrying and cutting, the expansion of Jodhpur town, the presence of military camps,[5] the requirements of workers at the site of the Rajasthan Canal being built towards the west, were factors that contributed towards widening the employment market and increasing the sources of income. No doubt there were landless labourers and small farmers still badly off with stagnant wages in real terms, but there were others who had known some improvement.

Let us return to Barli, 14 km. away from Jodhpur on the road to Jaisalmer. The village has a population of 1,879 (1981) out of which 332 are cultivators, 126 agricultural labourers, and 120 are involved in other jobs. The total work force is 578, out of which sixty-two are women.

The major problem was and remains that of drinking water. The pond dries up after a few months even if the rains are good. A few wells are located 2 or 3 km. away. In the other direction one walks the same distance as far as the nearest water pipe system. In Barli proper, all tests for finding ground water have failed until

[5] Defense and military camps were strengthened since the Indo-Pakistan war of 1965.

now. Sometimes the quarry owners arrange for a truck to bring water from Jodhpur.

We spent some time again with the Kumhars (potters by tradition). In 1979, they were busy for six months of the year making pots and earning Rs. 2.50-Rs. 3 per day. One family had a cow which gave 1.5 l. of milk a day which they consumed at home. They would also find work in construction or help in the quarries for Rs. 5 per day.

Today, the Kumhars save Rs. 10-12 per day (net) from their traditional occupation. From construction they earn Rs. 15, and Rs. 12 for stone quarrying and cutting, which means an increase in real wages as compared to 1979. Sohan Lal, one of the Kumhars, is proud of his bicycle which he has recently bought out of his savings. To save money, the Kumhars follow the custom, commonly found in Rajasthan of marrying several couples in a common ceremony. Despite this it costs them Rs. 2,000 to get a girl married.

Shivelam, Purohit, is a farmer who owns 3.2 ha. of land. His land bears *bajra* when the rains are good, but only part of the area is cultivated each year. He has a cow and is not too poor because he is able to find casual work in stone cutting for Rs. 12 per day.

His neighbour, a tailor, was behind his old sewing machine when we met him. He is quite happy because he now earns Rs. 25 per day, as against Rs. 8 in 1979.

Driving towards the old temples of Osian, we stopped in some villages where tubewells are used. Hukum Ram, a lively and shrewd farmer owns, with his brother, 9.3 ha. of land. The 9 HP diesel engine and the tubewell cost Rs. 50,000. They were able to get a loan of Rs. 15,000. It could supply more water if it was deeper than the present 21 m. On one-third of the land, Hukum Ram grows chilli (for seven months) which was being harvested when we met him. He makes a gross profit of Rs. 35,000 and a net profit of Rs. 15,000. He also grows wheat and mustard. He keeps some land free for his four cows to graze. Daily labourers earn Rs. 6 or Rs. 10 for harvesting the chilli. Several other farmers have a similar crop pattern thanks to the tubewells.

At Rampur (30 km. from Jodhpur) Persian wheels had been in use for a long time. Now they have become dry because, due to the boring of tubewells the water table has gone down, a phenomenon characteristic of the process of changing techniques (I observed it in Nilkanth in the 1970s). While one farmer complains, another has just completed a tubewell (130 m. deep) costing Rs. 100,000. He has lined his field channels with local material to save water. Next to his land, four brothers jointly own 16 ha., part of which is irrigated, enabling them to grow chilli and wheat as well.

On leaving Jodhpur towards Barmer one can see the road surrounded by 600 ha. of well irrigated crops. Fields of cabbage, tomato, carrot, chilli, wheat and mustard make us forget that we are in the desert. Tubewells (17 HP) lift the water from 80 m. The cost of digging and boring in this rocky terrain is Rs. 50,000. The pump, the motor, the electrical connections and field channels together cost Rs. 40,000. Water is distributed first through underground pipes and then through lined channels, which enables 15-20 per cent of the water to be saved.[6]

Yet, these costly investments (for a command area of 10 ha.) do not ensure an entirely reliable irrigation. When water is not replenished due to drought, one-fifth of the land is left fallow.

C.L.Bhati's father owns one of the tubewells. He himself is a good example of a well off, enterprising gentleman farmer. He cultivates 10 hectares of irrigated land with his 35 HP tractor. Wheat gives a very good yield of 4,000 kg./ha. Chilli, tomato and other vegetables cover a large part of the land. Here again, one can see that vegetable crops do not always guarantee large earnings. The price of chilli can vary from Re. 1 per kg. to Rs. 5. Besides, it requires a lot of farm manure and chemical fertilisers. Chilli and tomato can also be wiped out more easily by diseases.

In the *kharif*, C.L. Bhati grows *bajra* and *jowar* as fodder. His two brothers look after the dairy farm and their 100 cows, several of which are cross-bred. About 200 l. are sold everyday at Rs. 2.50 to Rs. 3 per l. (Not all the cows supply milk at the same time. They employ thirteen labourers on a daily basis at a wage of Rs. 10.). The fourth brother works as a contractor in Jodhpur.

A few kilometres further we visited the smaller farmers. Jeru Ram, a Patidar, and his two brothers are young, smart and hard working in the tradition of their caste. They have to manage — luckily they do not have large families to support — out of 1.3 ha. Their well (70 m. deep) relies on a 17 HP engine, necessary for

[6] A very clever, but rather costly device, not found elsewhere during this tour of India, except on field channels which had been lined by the government.

such a depth. Irrigation is not totally secure so that wheat yields vary between 1,200 and 3,000 kg./ha. depending on the supply of water. They also complain of damage caused by birds. The brothers keep a part of their land for chilli production which brings a much higher return than wheat. Their six cows supply 10 l. per day.

On the neighbouring farm five brothers live with their father. They own 4 ha. of land. The pattern of irrigation and cropping are similar to the other landholdings. Finally, we came across another landowner who, with the help of his tubewell, is developing fine orchards of citrus and mango trees. He also cultivates roses.

Such progress as has come about in the past ten years is nc doubt encouraging, but it also shows how the progress of agriculture differs from areas where there are no water constraints. Here, not only is the investment in digging a well much higher than in alluvial soil where one simply never sees a stone, but irrigation may be affected by drought, again a risk practically non-existent in alluvial basins. Third, the scope for irrigation is limited : a few kilometres away, we could see once again poor, unirrigated land.

Similar facts emerge from most areas of the district. In Osian block, around 200,000 ha. can be cultivated with reasonably good rains. Only 12,300 ha. are irrigated as a result of 2,850 wells-cum-pumpsets (10-35 HP). However, 500 are not in working order because of the drought conditions at the time, or because of maintenance difficulties. Power break-downs may also occur. Irrigated wheat yields 1,500 kg./ha., *bajra* a few quintals. Daily wages range from Rs. 7 to Rs. 11.

In Mathania, a big village of 7,000 inhabitants with a lively bazar and many workshops of all kinds, we visited some Julaha families, traditional weavers who were helped by the IRDP. They receive cotton and polyester thread from the officials. Working on their handlooms they sell the cloth to a government agency and earn Rs. 15 to Rs. 20 per day.

We move further up to Jat Khani, a village of 2,000 people. Here there are no fields of wheat, no wide open spaces red with drying chilli. Trees are rare, bushes are scattered on sandy soil. There are two wells supplying, for two hours a day, dirty water for drinking. Otherwise women go 3 km. away, three times a day to fetch water.

The government has started a water conservation project which employs 140 people earning Rs. 11 per day, half in kind, half in cash. It is worth noting that mostly women work there, which implies that the men have gone out for some other job, especially in or around Jodhpur. While the children were rather dirty — not much water is used for washing — the women were well dressed in typical Rajasthani fashion, a sight very different from the poor Chamar and Dusad women in Kalpi. The cattle were still surviving, but because of reduced fodder and water the cows would only give 2 l. of milk per day instead of 4 l.[7]

Jodhpur district

Jodhpur is,[8] like other districts in west Rajasthan, a large district of 22,720 sq. km. with a population which has practically doubled from 1961 (884,000 inhabitants) to 1981 (1.7 million). Between 1971 and 1981, the increase amounted to 44.8 per cent. This increase is surprisingly high and cannot be accounted for merely by the fact that family planning has had little impact here. I suspect that there has been a sizable under-enumeration during the 1971 Census.

Cultivated area varies a great deal, following the impact of the weather. *Bajra* covered from 401,000 ha. in 1976-77 to 550,000 ha. and up to 625,000 ha. during 1980-81, a year exceptionally good for crops. Production variations are even higher.

Average annual production of bajra in tonnes		*Yield kg./ha.*
1969-70 to 1973-74	153,000	230
1974-75 to 1979-80	88,000	164
1980-81 to 1984-85	98,000	179

The data for 1985-86 were not yet available during my visit, but production had been extremely low.

Being irrigated, though with the qualifications observed in the field, wheat was less erratic, although the impact of weather is clearly felt here unlike in the irrigated alluvial areas.

[7] Note that both yields here are quite high, compared to the yields in the Ganges basin where cows often give only 1l.

[8] Data supplied by the district authorities. See also Sankhya Ruprekha, Jodhpur *District Statistical Handbook*, Jodhpur Statistical Office, 1982, and CAZRI *Population, Land, Crops and Livestock Statistics of the Arid Zone of Rajasthan*, Jodhpur, 1983.

Average annual production of wheat in tonnes		*Sown area (ha.)*	*Yield kg./ha.*
1969-70 to 1973-74	28,000	26,200	1068
1974-75 to 1979-80	48,000	44,000	1090
1980-81 to 1984-85	38,000	33,000	1115

Various types of pulses cover an area ranging from 170,000 to 290,000 ha., with even greater output variations : the lowest being 10,000 t. in 1980-81, the highest being 37,000 t. in 1976-77. Mustard varies from 2,900 ha. (1977-78) to 17,300 ha. (1980-81) and from 1,340 t. to 10,700 t. for the same years.

All these data give us a clear idea of the enormous impact of the weather on production. Data for Jodhpur division (which includes the seven western-most districts) indicate very similar variations. The output of wheat in 1984-85 was exactly half that in 1979-80, which was a bumper crop. For *bajra* the peak year was 1973-74 with 2.2 million ha. sown, falling to 570,000 ha. in 1985-86 producing only 22,000 t.

Irrigation depends mostly on wells-cum-pumps or tubewells. The water level may fall during the growth of *rabi* crops like wheat. As a result, farmers stop irrigating part of their fields and concentrate on irrigating a smaller sown area. The plots which are no longer irrigated either yield a poor crop or no crop at all. It is quite obvious that the population-foodgrain ratio has deteriorated in the past twenty years, in spite of the lack of data or its accuracy, for the population, cultivated area and yields.[9] There is no indication of any deterioration in the climate. Such fluctuations as there are have been there for a long time. According to an old saying, in one century, there are twenty-seven good years, sixty-three intermediate, seven very bad and three disastrous. The new dimensions added to these basic problems are the large increases in population and in cattle, goats and sheep, which put an increasing pressure on the already fragile natural resources.

While the sown area is around one million hectares as against approximately 900,000 ha. in the late 1950s, irrigation has expanded:

Irrigated area (ha.)

1960–61	20,500	1978–79	43,000
1970–71	25,700	1985–86	56,000

[9] There must be a good deal of approximation about the cultivated area for, and production of *bajra* in view of such variations. Land registration data may also inaccurate.

Most of the area depends on wells and increasingly on tubewells. Further expansion seems unlikely as even now the water supply is not assured. The potential of surface water has been partially exhausted. Not only are there few rivers and even fewer possibilities for new reservoirs, in such hot areas 'nearly 40 per cent of the stored water is amenable to evaporation.'[10] The greatest hope lies with the Rajasthan Canal which is under construction (now renamed the Indira Gandhi Canal). This is a gigantic project, which starts from Harike Barrage on the Sutlej in Punjab. The main canal has already reached Jaisalmer and should end in Barmer. It provides water for both irrigation and drinking purposes.

Data on the Indira Gandhi Canal (Stages I and II)

Main canal	650 km. (up to Jaisalmer) not yet completed
Distribution system	7000 km. under construction
Cultivable command area	1.66 million ha.
Already completed	0.4 million ha.
Total estimated cost	11.25 billion rupees (1985 prices)
Expenditure incurred	4.5 billion rupees

Source: Indira Gandhi Canal Authority, Jaipur, January 1986.

The major part of Stages I and II should be completed by 1994-95. All canals have been lined to avoid water loss through seepage, except field channels fitted with outlets for every 6.25 ha. In several areas, gravity canals cannot be dug because of the land profile, which is why lift canals (i.e., branches with large lift pumps) are needed.

A branch canal of 210 km. is being constructed, cutting through Jodhpur district from north to south. On each side of the canal there will be a 3 km. broad belt of irrigated land. It will enable the light irrigation of *bajra* which could boost yields to over 1,000 kg./ha. Fodder, wheat and vegetables could also be grown. The project is even more important as drinking water will be made available to the rapidly rising population of Jodhpur and the villages along the canal. It is estimated that drinking water will reach Jodhpur city by 1989.

Though highly welcome, such a project will not alleviate all difficulties. It could add roughly 100,000 ha. to irrigated land, so that

[10] CAZRI, *Progress in Arid Zone Research*, (Jodhpur, January 1983), p. 17.

the whole district would then have around 150,000 irrigated ha. out of one million cultivated hectares. Only the vulnerability of the district would thus be reduced.

In view of these limitations, one can envisage a more efficient use of irrigation, thanks to reduced losses. This is already being done by lining the canals (a very expensive venture) and, as seen in some villages, lining the field channels of the tubewells. More modern methods of saving such as the sprinkler and drip have also been tried out. In 1979, I saw these on farms belonging to Rajput landlords or very modern gentlemen farmers. The trouble is that such sophisticated techniques are costly, with the result that there is little hope of their becoming widespread. Between my two visits, in fact, there had been little progress.

In view of the limited pattern of irrigation, one has to look for other ways out : soil improvement and better preservation of soil moisture, for example. It does not seem that the desert, or semi-desert, is really expanding, but within the present area the soil has very definitely deteriorated. More marginal land has been opened to the plough, and erosion and over-grazing have caused grazing grounds to deteriorate and shrink. The inevitable process of desertification is accelerated when, under land reform programmes, large areas of common property, 'mostly unsuited for cultivation were distributed as crop lands to the landless.'[11]

Evolution of the twelve desert or semi-desert districts of Rajasthan

	1901	*1981*
Population (million)	10.3	34.1
	1956-57	*1977-78*
Fallow land (million ha.)	4.4	3.49
Net area sown (million ha.)	7.76	9.78
Cattle (million heads)	4.14	4.27*
Sheep (million heads)	4.89	6.86
Goats (million heads)	3.75	6.43

*Cattle peaked at 4.94 million in 1966 then fell to 3.7 million in 1972 due to drought.

Source: CAZRI, *op. cit.*

Jodhpur district fits well into these broad trends.

Tractors have known a rapid expansion and about 60 per cent

[11] N.S. Jodha, Population Growth and the Decline of Common Property Resources in Rajasthan, *Population and Development Review*, no 2, June 1985.

of the sown area relies on them. Extensive cultivation on wide fields makes the use of tractors feasible and, in addition, quick tractor ploughing helps retain soil moisture. On the other hand, tractors stimulate reclamation of marginal land and lead to soil erosion through winds. Do the advantages of tractors outweigh their defects? This may be so because tractors have the added advantage that they are used for transportation. The question, however, remains a complex one.

Field bunding, as recommended by CAZRI may help. *Bajra* yields may increase by 10 to 12 per cent.[12] Some amount of storage, however small, of run-off water can also increase crop yields or save them from loss when rains fail after the first showers.

Afforestation and grass development are key items in the strategy against desertification. Stabilising sand dunes, as we saw in 1979, involves fencing with barbed wire (at a cost of Rs. 2,000-3,000 per ha., at 1979 prices) planting of 550 trees per ha., and in between, seeds of grass. After every ten years, farmers collect 5 t. of wood. In between they get 1 t. of fodder. Such schemes and methods are no doubt interesting, but they are both expensive and time consuming. As it is, new damages to the environment are proceeding faster than the remedies.

After revisiting CAZRI, I was taken to the Jodhpur Agricultural Centre affiliated to the university. Here one appreciated again the complexity of and the time needed for research on dry lands and the problem of light irrigation. For instance, castor oil increases its yield by 10-15 per cent if the first watering is given after thirty days instead of forty-five days as is the normal practice. New varieties of mustard can yield up to 1,500 kg./ha., as against the more usual yield of about 750-800 kg./ha. On the other hand, some new pulses failed because of the extreme heat. High yield *bajra* may give a 30 per cent higher yield even without chemical fertilisers, but, at the moment only one new variety is available which is quite vulnerable to diseases. Some chemical products are being experimented with, which, when used on the soil, can increase the amount of moisture conservation. Pulses and oilseeds, so much in demand all over India, are progressing slowly in research centres, and so far new varieties have not led to such striking increases in yield, compared to what we have seen with wheat and, to a lesser extent, with rice.

[12] CAZRI, *op. cit.*, p. 26.

We must also not forget that this research remains confined to irrigated crops which cover a small part of the land. As for *bajra*, in a very good year it can go up to 500-1,000 kg./ha. With poor rains it falls to 0 to 200 kg. and the cropped area also shrinks. It is difficult to visualise how such variations can be converted into a long term upward trend, especially when taking into consideration that marginal land already accounts for 20 per cent of the cultivated area.

Although animal husbandry endangers the environment through over-grazing, at least in certain areas, it remains an important asset for the farmers. The commercialisation of milk production through a network of cooperatives and trucks collecting milk from villages, was already visible in 1979. At the time, 30,000 l. of milk were transported every day to Delhi. Today, collecting and chilling units continue to expand but with ups and downs. Under conditions of a normal rainfall, grass and small bushes supply enough fodder and we have seen that cows give more milk here than in the Ganges basin, often 2-3 l. as opposed to 1 l. But when the rains fail, as in 1985, the situation becomes precarious. Herdsmen try and migrate with their cattle to better areas. Some cattle die, some are saved through relief measures — which are unavoidably limited — with trucks bringing fodder from other areas.

Another asset in Jodhpur district is its high rate of urbanisation.

Population of Jodhpur town

1931	95,000
1951	181,000
1981	506,000

The percentage of urban population keeps on rising : 32.2 in 1971, 34.7 in 1981. By that time, out of 510,229 working persons, cultivators amounted to 300,000 and agricultural workers to 25,000 as against 185,000 in household industry, manufacturing, services and other works.

The city of Jodhpur has several medium and small industries and flourishing handicrafts (textiles and leather goods) in the villages. The large military base also creates job opportunities here.

The construction work at the Indira Gandhi Canal site has offered an outlet for some time now, a trend which can only expand with the new works on the canal going through Jodhpur district.

The progress in infrastructure with the expansion of roads and electricity, has made the overall process of growth and diversification of the economy easier. To this may be added foreign tourism though it still plays a minor role.[13]

Maybe the traditional business background of the so-called Marwaris should also be taken into account. Although the most enterprising ones have left Marwar (the ancient name of the princely state of Jodhpur) to start industries all over India, especially in Calcutta and Bombay, several local traders and moneylenders are now turning to industry.[14]

Social issues

All these factors are indicative of how Jodhpur which, in the British days, was much less developed than Jaipur, has known a sizable development which has also had an impact on small farmers. In spite of such severe natural constraints, one does not come across very acute poverty.

	Daily wages (Rs.)
1979	2.50 – 3.00 – 4.00 – 5.00
1986	5.00 – 7.00 – 10.00 – 12.00

The lowest wages are found in relatively remote and poor villages, the others in less isolated areas, following a pattern observed in other districts. On the whole, wages in real terms have increased.

While there are few landless labourers (see later) and landholdings are much larger than in other parts of the country, we must not forget that double-cropping is rare and yields per rainfed crop extremely low. Out of 152,000 landholdings, nearly two-thirds are above 4 ha., out of which 43,000 holdings are above 10 ha. (1976-77).

[13] Many foreigners visit the fascinating city of Jaisalmer in the heart of the desert. They come from Jodhpur, itself a very ancient city.

[14] Out-migrations of Marwaris started in the latter part of the last century. Marwari has now become a general word to refer to a banya (trader) from any part of Rajasthan.

Although one comes across new drinking water schemes, often equipped with pipes, the situation is still difficult even in normal years.

A number of projects are devoted to the poor. During my first visit in 1979, loans-cum-subsidies were available in favour of small farmers, especially for animal husbandry. They did not prove very successful, because, in case of drought the first to go without fodder are the poor. A loan-cum-subsidy could provide thirty sheep, which in three to four years could multiply to a hundred. What in fact happens is that a person is left with a dozen heads only. Similar problems occur with milch animals. That is why today the emphasis is more towards small business or industry, like the weavers referred to earlier, or providing help for embroidery, making carpets and leather shoes. The problem here is the complexity of the operations. One needs an efficient marketing cooperative, adequate supply of inputs and proper demand. It is much easier to open a small shop instead. However, I cannot be certain of the final scope of such projects.

During my visit in 1986, relief work was in operation because of the drought. People were being given some employment and money to tide over the difficult times. Trucks with water and sometimes with fodder were also trying to cope with at least parts of the most badly affected areas. It is always difficult, unless one starts a detailed enquiry, to pass judgement on these relief operations. On the whole my impressions were positive and were strengthened after meeting several district officers : the collector and other IAS officers in charge of the DRDA and the Desert Development Programme as well as agricultural officers. In Jodhpur district also, I gathered the same feeling, as evidenced elsewhere, of better development administration compared to the past decade or so.

Jodhpur district represents several complex challenges. Here the ecology and development are very closely inter-related. Until now the deterioration of the eco-system had not prevented a sizable development, including some progress for the poor. Yet, the future is worrying. How far and how can the process of economic diversification be pushed in view of the halting progress of agriculture? What will be the final impact of the branch of the Indira Gandhi Canal in the district? Last but not the least, is the question of family planning. Jodhpur, like the rest of Rajasthan, has made very little progress and may pay dearly for it if drastic changes are not introduced.

Summing up

It is time, after this long trek through this diverse country, to summarise our main findings. Even in 1978-79 one could observe techno-economic and socio-political changes in practically all the regions visited, including remote tribes in the forests of Orissa. At the same time, the pace of change varied enormously: it was considerable in certain districts yet, slow in others. These same observations were repeated seven years later, with one qualification: the trends perceived in those days were now, more obvious.

Population pressure is increasing everywhere, leading to a further division of holdings. More and more landowners have to manage on a shrinking sown area and the number of landless labourers continues to rise. While family planning is needed more urgently now than ever before, one does not come across striking changes, compared to our earlier surveys. The sluggish villages remain slack, the progressive villages remain ahead of the others as before. There are areas where the agricultural output and the global process of rural development are moving fast enough in order to boost agricultural wages and employment opportunities, so that a number of the poor do benefit from economic growth.

In other parts of the country, economic forces are moving too slowly to be able to positively affect the lot of the poor. Their condition may not have deteriorated, but, in many cases, it is hardly improving.

While new employment opportunities, including the proliferation of odd secondary jobs, cannot be measured accurately, the trends in daily wages are more easily assessed. In 1985-86, in advanced areas, they would range between Rs. 10 to Rs. 12 if not more, and about 10-12 kg. of grain at the time of harvest. In slow moving areas, cash wages would be around Rs. 4.50 and Rs. 6, and the share of the harvest would amount to 3 to 5 kg.of grain. In advanced areas there has been some increase in cash wages in real terms as compared to 1978-79 when it was Rs. 5 to Rs. 6. In the sluggish areas there has been very little or no increase in real terms: wages stood at Rs. 2.50 to Rs. 3.50 in 1978-79.

We found, as in our previous survey, that practically all farmers in progressive areas would resort to new inputs but would sometimes use more moderate amounts of chemical fertilisers or pesticides than farmers elsewhere.

As far as special schemes for the poor are concerned, one gets a mixed picture: definite success stories, but also a number of complaints about all kinds of leakages and malpractices.

The district administration made a better impression than on our previous visits. Officers of the Indian Administrative Service are often of a high calibre. Compared to our previous surveys, several of the other district officers as well are more knowledgeable, have a better understanding of local issues and are more committed to their task. At the same time, however, farmers are more vocal than ever about corruption and other malpractices.

Several techno-economic problems have grown in magnitude and complexity. The shortage of electricity has become a major constraint everywhere, but in varying degrees. The same can be said of the operation and maintenance of canal irrigation systems and of drainage problems.

In advanced districts like Bulandshahr, further increase in production will be very demanding in terms of efficiency, adequate supply of inputs and continuous research. In the eastern plains, on the other hand, considerable progress can be achieved with the 'muddle through system.' As far as peninsular India is concerned, the efforts to increase rainfed crops will take time and the results will remain much below the breakthrough achieved in the north-west and — though more limited — in the south-eastern deltas.

Local politics has not improved in most areas. In Bihar, one can even speak of a further deterioration with all manner of adverse effects on the socio-economic situation.

The problem of the new generation of farmers, the young and educated sons of medium and upper landowners, as seen in Bulandshahr, is not particular to that area alone. It plays a role in Punjab, but it may be more acute in progressive areas than elsewhere. Other social issues, like increasing expenditure on marriages lead to painful and sometimes even tragic situations.[15] Caste relations also throw up all kinds of situations : some very bad, others indicating that there has been some improvement.

[15] *India Today,* 15 June 1986, reports cases of infanticide among young girls of the Kallar caste of Madurai district, to avoid future marriage expenditure.

Chapter Thirteen

The All-India Perspective

After visiting villages and districts, interviewing farmers and local officials, it is time to move on to the all-India perspective, in order to see how these field observations fit into the broader framework.

Agriculture and population

Foodgrain statistics and population censuses point to a clear improvement in the overall food situation. After facing a long deficit in cereals, India has become self-sufficient and enjoys comfortable stocks even in the face of severe natural calamities.

	Population (million persons)	*Foodgrain output million t.*
1949-50[1]	355†	55
1960-61	439	82*
1970-71	548	108.4*
1980-81	685	129.6*
1983-84	730†	152.4*
1984-85	745†	145.5
1985-86	760†	150.5
1986-87	775	151†

* very good monsoon. Population + 118 per cent Foodgrain + 175 per cent

† Figures for three years are based on estimates. For the remaining years figures are based on census data.

Foodgrain imports stood at around 3 million t. in the late 1940s and early 1950s. After a temporary fall they rose in the 1960s up to a peak of 10 million t. in 1966. Then they fell again, were discontinued in the early 1970s and came up again later. During the period 1977-81, imports were discontinued again, to reappear during 1982-84 (2.7 million t. per year). These last imports were merely to replenish government silos rather than to cope with an

[1] 1949-50 is the base year, because 1950-51 was an exceptionally bad year with an output of 51 million t.

immediate shortage. Since 1985, imports have been discontinued once again and replaced by exports of a few hundred thousand tonnes of foodgrain per year.

Despite the gloomy and perfectly ill-founded forecasts or 'prophecies' about India's future disasters, which were fashionable in the 1960s, the country is no longer exposed to real famines. And, in case of a bad monsoon, India can first rely on its stocks (29 million t. of grain in June 1985; 23.6 million t. in December 1986), instead of resorting to imports or worse still, to international charity as was the case during the dramatic drought years of 1965 and 1966.

Increase in crop production (million tonnes)

	1949-50	1970-71	1983-84	1984-85	1985-86
Oilseeds	5.2	9.3	12.7	12.95	11.15
Cotton lint	0.47	0.86	1.10	1.44	1.46
Sugarcane	61.0	130.0	174.0	174.1	171.7
Jute	0.72	1.13	1.14	1.18	1.97
Tea	0.27	0.42	0.64	0.65	NA

The period starting in 1950 can be divided in two. Until the early 1960s, production increased mostly within the framework of traditional means and techniques—land reclamation, additional irrigation canals and wells. New inputs began to appear later. Tubewells and pumpsets expanded fast (there were already some in use even during the early phase). The consumption of chemical fertilisers rose sharply, accompanying the progress of irrigation and new seeds, especially of cereals. Pesticides were introduced. Although tractors do not play a major role in increasing the yields of crops, they may affect production. Tractors, which were exceedingly rare in the 1950s, are fairly widespread today, at least in advanced districts, and in Rajasthan. However, at the all-India level, bullocks remain the predominant means of ploughing. In 1985, there were about 500,000 tractors in use.[2]

The new inputs led to increase in yields per crop and more multi-cropping, while land reclamation possibilities have been exhausted for several years.

[2] Government of India, *Seventh Five Year Plan 1985-90* (New Delhi, Planning Commission, 1985), vol. 1, p. 2.

Cultivated land (million hectares)

	Net	Gross
1950-51	119	139
1979-80	142	173
1983-84	143	177

Actual irrigated areas (gross)

1950-51	1960-61	1970-71	1983-84	1984-85	1985-86
22.6	27.9	40	58.6	60.5	62.3

The progress of tubewells and pumpsets on open wells has been particularly striking. The consumption of chemical fertilisers, both locally produced and imported,[3] was practically unheard of except in some tea plantations and other estates.

Consumption of chemical fertilisers

1950-51	62,000	in tonnes of (nutrients)
1960-61	292,000	
1980-81	5,516,000	
1983-84	7,709,000	
1985-86	8,740,000	

The progress in the use of pesticides has been more modest and estimates on their consumption are not too reliable. But 50,000 t. in 1984-85 and 66,000 t. in 1985-86 can be a rough estimate.

All these data are no doubt interesting, but they obscure such enormous differences that they can be misleading. Taking food-grain first, the progress has been really spectacular for wheat, less so for rice, and moderate for the other crops.

	1949-50	*1983-84*	*1984-85*	*1985-86*
		(million t.)		
Wheat	6.8	45.5	44.1	46.9
Rice	25.1	60.1	58.3	64.2
Coarse grains*	18.7	33.9	31.2	27.0
Pulses	10.0	12.9	12.0	13.0

*Mostly *jowar, bajra* and maïze.

The impact of drought in peninsular India explains the fall in coarse grains after the bumper crops of 1983-84. The progress of

[3] Local production : 1980-81 3 million t.
1985-86 5.7 million t.

wheat has remained steady with the help of irrigation. Irrigation and generally good rains in the main paddy growing areas, explain the latest progress in rice.

These differences are linked to the impact of irrigation, soils, and use of chemical fertilisers. Out of 23-24 million ha. of wheat, 16 million or more are irrigated; out of 40-42 million ha. of rice, only 18 million benefit from irrigation. Usually irrigated crops use more chemical fertilisers. Coarse grains are mostly non-irrigated, grown on soils which are often poor. Pulses can be grown on both poor and rich soils.

The development of infrastructure has contributed to overall rural development:

	1951	*1985*
Number of villages electrified	3,061	369,000
Roads	400,000 (km.)	1,770,000 (km.)

In 1985, 64 per cent of the villages were electrified. The network of roads connected 64 per cent of all villages, but about two-thirds of these roads are only fair weather roads, i.e., subject to temporary damage by heavy rains.

Population growth and land-man ratio

Population growth (state-wise) presents interesting variations between the decades 1961-71 and 1971-81.

	*Area (As on 31-3-1962)**	*Population*		
		1981	*1971*	*1961*
	'000 km	*'000s*		
States:				
Andhra Pradesh	275	53,550	43,503	35,983
Assam	78	19,897†	14,625	10,837
Bihar	174	69,915	56,353	46,447
Gujarat	196	34,086	26,697	20,633
Haryana	44	12,923	10,037	7,591
Himachal Pradesh	56	4,281	3,460	2,812

(Contd.)

(Contd.)

	*Area (As on 31-3-1962)**	*Population*		
		1981	*1971*	*1961*
	'000 km	*'000s*		
States: (Contd.)				
Jammu & Kashmir	222	5,987	4,617	3,561
Karnataka	192	37,136	29,299	23,587
Kerala	39	25,454	21,347	16,904
Madhya Pradesh	443	52,179	41,654	32,372
Maharashtra	306	62,784	50,412	39,554
Manipur	22	1,421	1,073	780
Meghalaya	22	1,336	1,012	769
Nagaland	17	775	516	369
Orissa	156	26,370	21,945	17,549
Punjab	50	16,789	13,551	11,135
Rajasthan	342	34,262	25,766	20,156
Sikkim	7	316	210	162
Tamil Nadu	130	48,408	41,199	33,687
Tripura	10	2,053	1,556	1,142
Uttar Pradesh	294	110,862	88,341	73,755
West Bengal	89	54,531	44,312	34,926
Union Territories:				
Andaman & Nicobar Islands	8	189	115	64
Arunachal Pradesh	84	632	468	337
Chandigarh	0.1	452	257	120
Dadra & Nagar Haveli	0.5	104	74	58
Delhi	1.5	6,220	4,066	2,659
Goa, Daman & Diu	3.8	1,087	858	627
Lakshadweep	0.03	40	32	24
Mizoram	21	494	332	266
Pondicherry	0.5	604	472	369
All India	3,287	685,185	548,160	439,235

* Provisional.

† Official projection, but not Census estimate.

It is a matter of concern that the population is increasing very fast in already heavily populated states like Uttar Pradesh and Bihar, or in Rajasthan where the eco-system is so hard pressed. On the other hand, in several states, the growth rate is falling as in Kerala, Tamil Nadu, Gujarat, Maharashtra, West Bengal and

	Decennial growth of population (%)	
	1971-81	1961-71
Andhra Pradesh	23.1	20.9
Assam	36.1*	35.0
Bihar	24.1	21.3
Gujarat	27.7	29.4
Haryana	28.8	32.2
Himachal Pradesh	23.7	23.0
Jammu & Kashmir	29.7	29.7
Karnataka	26.8	24.2
Kerala	19.2	26.3
Madhya Pradesh	25.3	28.7
Maharashtra	24.5	27.5
Manipur	32.5	37.5
Meghalaya	32.0	31.5
Nagaland	50.1	39.9
Orissa	20.2	25.1
Punjab	23.9	21.7
Rajasthan	33.0	27.8
Sikkim	50.8	29.4
Tamil Nadu	17.5	22.3
Tripura	31.9	36.3
Uttar Pradesh	25.5	19.8
West Bengal	23.2	26.9
Delhi	53.0	52.0
All India (incl. others)	25.0	24.8

*Official projection, but not Census estimate

Madhya Pradesh. Family planning must be one of the main reasons for such a reduction, since most of these states[4] have followed a rather strong policy in this matter. Besides, these states have a relatively high rate of urbanisation which makes the expansion of family planning easier. The case of Orissa is puzzling. Family planning is not as advanced here and the urban population amounts to merely 12 per cent (1981). Has there been an increase in mortality during the decade 1971-81? According to

[4] The case of Madhya Pradesh however is not clear to me. Family planning campaigns have not gone very far and the urbanisation rate amounts to only 20 per cent.

the latest estimates of the Registrar General's office, the all-India birth rate was 32.7 per thousand in 1985 (33.9 per thousand in 1984), and the death rate was 11.7 per thousand in 1985 (12.6 per thousand in 1984).[5]

The rate of urbanisation is rising too slowly. It has risen to only 23.3 per cent in 1981, as against 20.2 per cent in 1971 and 18.3 per cent in 1961. As a result, the population pressure on land is getting heavier and heavier.

Even rough statistics on landholdings express the growing division of land, mostly through inheritance. According to Hindu tradition each son gets an equal part of his father's land.

	1970-71	*1980-81*
Marginal (below 1 ha.)	39.7	50.6
Small (1 – 2 ha.)	13.4	16.1
Semi-Medium (2-4)	10.7	12.5
Medium (4 – 10 ha.)	7.9	8.1
Large (10 ha. and above)	2.8	2.2
	74.4	89.5

Source: Government of India, *Agricultural Census* 1971 and 1981 (New Delhi, Ministry of Agriculture).

In 1980-81, 26 per cent of the total area was operated by marginal and small farmers, accounting for about 75 per cent of the total number of holdings. The all-India break up repeats itself in most states with slight differences. One striking exception is Punjab where the number of small holdings is decreasing.

The trouble with these data is that they do not take into account the productivity of the land. In an advanced district today, a farmer who owns even slightly below one irrigated ha. is not 'marginal' and can live reasonably well, which is not the case with a cultivator operating 3 ha. of rainfed land. As for landless and very small landholders, their number is on the increase.

Total working population (million)	
1971	184
1981	222

The percentage of cultivators and agricultural labourers is

[5] *Overseas Hindustan Times*, 14 February, 1987.

falling, but very slowly, so that the absolute figures keep on rising. Although there are state-wise or area-wise differences, they are not striking. The percentage of landless labourers to cultivators tends to be higher, and this is not new, especially in paddy-growing areas.

Percentage distribution of working population

	Total population		*Rural population only*	
	1971	*1981*	*1971*	*1981*
Cultivators	45.90	43.70	55.73	55.16
Agricultural labourers	21.54	19.56	29.61	24.00
Total	67.44	63.26	81.34	79.16
Other sectors	32.56	34.74	18.66	20.84

Source: J. Krishnamurty, Changes in Indian Working Force, *Economic and Political Weekly*, 15 December, 1984.

All these census figures must not be taken at their face value. At the field level we have seen how the poorer a man, the more difficult it is to assess his income and his various jobs. Particularly in prosperous areas, there may be a number of people classified as agricultural labourers and yet engaged, to a large extent, in miscellaneous activities.

With this caveat in mind, the broad trends that emerge from these tables remain a matter of very grave concern. At the all-India (rural and urban) level the changes in the employment pattern have been slow. Even within rural India, the process of diversification is proceeding slowly. Yet, we have seen how crucial it is to accelerate it in order to reduce poverty through more employment opportunities outside agriculture, not only in towns, but also in the villages itself in order to reduce the proliferation of urban slums.

Field observations of trends in family planning are confirmed at the all-India level. The decline in fertility has to be much more rapid though some states do better than others. Regardless of whether we view the problem of population from the dimension of food, employment or land, the conclusion remains the same: there is an urgent need to reduce the birth rate more rapidly.

The need for a regional approach

Let us return to our three broad categories of regions: advanced areas, potentially rich but still very poor zones, and, areas which face severe natural constraints. Though there have always been substantial regional differences in yields and output, they have increased considerably with the advent of new inputs in the 1960s.

Between 1970-71 and 1983-84 (two excellent monsoons) foodgrain production rose from 108.4 million t. to 152.4 million t. i.e., an increase of 44 million t. Of this, 23.4 million t. were supplied by Punjab (7.5 million t.), Haryana (1.8 million t.), UP (9.6 million t., the bulk supplied by the western districts), and, Andhra Pradesh (4.5 million t.) In Tamil Nadu, production has been levelling off since the late 1970s. This is due primarily to the very slow progress of irrigation.

In several states, the very good monsoon of 1983-84[6] was the major factor responsible for the growth, unlike in advanced areas, where new inputs played a larger role. In Assam, foodgrain output fluctuated at around 2.7 million t. of foodgrain (mostly rice) for the three years 1982-83 to 1984-85, to reach 3.03 million t. in 1985-86. In Bihar, between two bumper crops (1978-79 and 1984-85) there was practically no difference: 9.77 and 10.3 million t. but good rains in 1985-86 pushed foodgrain production to 11.2 million t. In Orissa the trend was better thanks to some progress in the coastal districts. Orissa produced 5.1 million t. in 1970-71, 6 million in 1980-81, 7.02 in 1983-84, but 6.7 million t. in 1985-86 due to a fall in pulses. In West Bengal, fluctuations were very sharp from year to year. The two years 1983-84 and 1984-85 each saw a production of 9.2 million t. against 9 million t. in the previous peak year of 1977-78 and only 8.8 million in 1985-86.

Fluctuations due to the weather are often more acute in the Deccan. After three consecutive drought years (1970-71, 1971-72, 1972-73), the weather improved in Maharashtra. However, after the peak year of 1977-78 (10.45 million t.), there was an increase only in 1983-84, when production rose to 10.95 million t. Then followed another series of drought years, foodgrain production falling to 9.7 million t. in 1984-85, 8.8 million in 1985-86

[6] 1982-83 suffered from a rather poor monsoon. As a result, foodgrain output fell to 129 million t. from 133 million t. in 1981-82.

and below 8 million t. in 1986-87. Data for Karnataka are not encouraging either, with a falling production in the two years following the bumper crops of 1983-84. They are a little more encouraging for Madhya Pradesh. The upward trend is more significant in Gujarat, until the disastrous drought of 1985-86 which cut production by half compared to 1983-84. In Rajasthan, fluctuations are particularly severe, especially in the western areas. Production has oscillated between 5 to 7 million t. for the past sixteen years, with a peak of 8.84 million t. in 1970-71 and another at 10.07 million t. in 1983-84.

To fully explain all these variations and differences is beyond the scope of this book and the ability of its author. Some factors, such as techno-economic data, geographical conditions, soils and climate are easy to assess. Other factors are much more complex, involving history, population, socio-political problems and caste class relations for which I will only attempt *some* explanation.

The areas in the plains

The differences observed in the field fit in well with more widespread conditions. In the early 1960s, good irrigated traditional agriculture would yield 1,200 – 1,500 kg./ha. of wheat or clean rice (1,800 – 2,250 kg./ha.of paddy). Bulandshahr illustrates the breakthrough in wheat, common to most western UP districts. The doab (region between the Ganges and the Yamuna) nevertheless falls behind Haryana and Punjab. Yields of irrigated wheat in Punjab have crossed the 3,000 kg./ha. average for the whole state (western UP stands at around 2,500 kg./ha.) Second, unlike in western UP, in both Haryana and Punjab, rice has replaced low yield maize, *jowar* and *bajra* in the *kharif*. Punjab's average yield amounts to 3,000 kg./ha. (irrigated) of clean rice or 4,500 kg./ha. of paddy, which is much higher than the average yields of coastal Andhra and Tamil Nadu.[7] Haryana stands at around 2,600-2,700 kg./ha. of clean rice. However, the irrigated areas of the two southern states have done quite well, with yields frequently around 2,000-2,500 kg./ha. of rice or 3,000-3,700 kg./ha. of

[7] We have quite frequently seen farmers in the South gathering around 3,000 kg./ha. of clean rice, but others get less. When we take average data for rather large areas, they fall below that of certain villages surveyed.

paddy. Irrigation, new seeds and chemical fertilisers have played a major role in such increases. Pesticides, especially for monsoon crops, have been introduced.

Coming to the eastern plains, the majority of paddy fields (the main foodgrain) remain rainfed with all the limitations this implies. New dwarf varieties are excluded from wide tracts submerged by too much water. The use of fertilisers is negligible. In Bihar, irrigated rice (state average) reaches 2,000 kg./ha. (3,000 kg./ha. of paddy) and rainfed rice is below 1,000 kg./ha. In West Bengal the main monsoon crop *(aman)* in the triennial ending in 1980-81 amounted to 1,310 kg./ha. of clean rice, and fell to 900 kg./ha. in the bad year 1981-82. However, *boro* rice (irrigated) harvested in March-April amounted to 2,700 kg./ha. The situation as regards Orissa fluctuated in the early 1980s between 750 and 950 kg./ha. of clean rice, mostly unirrigated (1,100 to 1,500 kg./ha. of paddy).[8] In Assam one finds yields between 1,000 and 1,200 kg./ha. of unirrigated rice.

	Percentage of irrigated area to cropped area 1980-81			*Consumption of chemical fertilisers kg/gross ha. (nutrients) 1980-81*
	Wheat	*Rice*	*All crops*	
Punjab	91	98	86	127
Haryana	93	97	61	47
Western UP	90	–	75	n.a.
Andhra Pradesh	–	94	35	53
Tamil Nadu	–	92	51	59
East UP[9]	70	40-50	56	64
Bihar	72	34	33	19
West Bengal	40	29	20	34
Assam	–	34	16	5
Orissa	–	28	20	10

Sources: Government of India, *Area and Production of Principal Crops in India* (New Delhi, Ministry of Agriculture, Directorate of Economics and Statistics, 1981-84). Reserve Bank of India, *op.cit.*, vol. 2. Government of India, *Draft Seventh Five Year Plan*, vol. 1, p. 438, Uttar Pradesh, p. 14.

[8] For all these data, See Reserve Bank of India, *Report of the Committee on Agricultural Productivity in Eastern India* (Bombay, 1984), vol.2.

[9] Central UP would deserve a detailed study for which I do not have adequate information. Its growth rate in foodgrain (58 per cent for 1960-61 to 1981-82) falls behind eastern UP (78 per cent) and western UP (116 per cent). Its irrigated area amounts to 52 per cent of the sown area. For data on UP *Seventh Five Year Plan*, p. 111-114.

The above data give us rough estimates of the gap between already fairly advanced areas and potentially rich but still very poor regions in this country.

These differences were not seriously altered during the Sixth Plan. In 1985-86, Punjab reached 150 kg./gross ha., Tamil Nadu 100 kg./gross ha. There has also been a substantial increase in Haryana, UP and Andhra Pradesh, but Bihar and particularly Orissa, are 'lagging far behind' the national average of 50 kg./ gross ha. in 1985-86 (*Economic Survey, 1986-87*, p. 13).

These data require serious qualifications. For chemical fertilisers, the actual consumption is much above the state average for coastal Andhra and coastal Tamil Nadu. In both states, the wide plateau areas have a very low consumption. The figure for Haryana too looks surprisingly low. In terms of irrigation, considerable differences can be noted. In the north-west, irrigation is relatively adequate and reliable. Canals here are less affected by drought than in the south. Tubewells play a prominent role and are not vulnerable to drought. In the eastern plains, tubewells are less widespread, although they are now progressing rather fast in eastern UP. The supply of electricity, especially in Bihar is worse than in the north-west.

With reference to the south, canals do play a major role, not in all coastal districts, but in several of them. Since their water supply depends only on the monsoon (unlike for the rivers coming from the Himalayas) irrigation is not immune from drought and the canals do not allow perennial irrigation. Second, the sizable areas of Andhra and Tamil Nadu depending on tanks are even more vulnerable to the vagaries of the monsoon. There is another problem, that of wasteful field to field irrigation in the south, compared to the more efficient system of the canals in the north-west from which outlets bring water to smaller plots of land.

Finally, we cannot give the break up of irrigated land between alluvial and plateau areas as found in Bihar, Orissa, Andhra and Tamil Nadu.

While the operation and maintenance of canals need considerable improvement in the south-east deltas, the situation is worse in the eastern plains.

In spite of their slow growth, the plains of eastern India are no longer static. The whole of eastern UP, an area long considered unpromising, is now awakening. The total irrigated area has risen from 2.3 to 2.8 million hectares between 1971 and 1981, mostly through the expansion of tubewells (in 1981 1.7 million ha. had been irrigated). Wheat, the main beneficiary of irrigation, rose from 0.6 million t. in 1960-61 to 3.9 million t. in 1981-82, with an average yield of 1,500 kg./ha., as against 740 kg./ha. in 1960-61. Unfortunately, the progress in rice (35 per cent of the total cropped area) has been much less spectacular and much more erratic than wheat, due to less irrigation. From 650 kg./ha. of clean rice in 1960-61, it reached nearly 1,000 kg./ha. in 1981-82. It had, however, fallen to 400 kg./ha. during the severe drought in 1979-80.

In Bihar and West Bengal, the progress of wheat must be emphasised. In Bihar the area under wheat doubled in twenty years to reach 1.8 million ha. in 1980-81 (average yield 1,560 kg./ha.). In West Bengal the area remains small: around half a million hectares, but the yield is high with 2,600 kg./ha. In both cases unfortunately, at the state level the success of wheat has had a limited impact in view of the dominance of rice.[10]

Non-agricultural factors enable us to assess other regional differences.

	Surface road length per 100 sq.km. of area	*Percentage of villages electrified*
Punjab	71	100
Haryana	45	100
UP	24	47
Western UP	not available	63
Andhra	23	76
Tamil Nadu	71	99
Eastern UP	not available	51
Bihar	16	43
West Bengal	28	46
Assam	25	not available but estimated as being low.
Orissa	11	45

Source: Government of India, *Draft Seventh Five Year Plan*, Uttar Pradesh, p. 14-15

In western UP the area irrigated by power-operated private tubewells amounts to 1.3 million ha. versus 0.56 million ha. in eastern UP.[11]

[10] For all data see, Reserve Bank of India, *op. cit.*, vol. 2.

[11] *A Study of Electric Power Consumption in Agriculture in Uttar Pradesh*, Lucknow, 1984 (Report of a special task force under the Agricultural Production Commissioner).

In spite of the lack of data on surface roads, area-wise in UP, there is little doubt that, again, western UP is more advanced than the rest of the state.

For Andhra, it would be desirable to have independent data for the coastal areas and the interiors, but it is obvious that the coastal districts are much better equipped (with roads and electricity) than the interior.

Even if the system of roads and the electrical network are more developed in the plains of Bihar and Orissa than in plateaus the overall gaps are striking compared to advanced regions. Besides, one must consider how the state electricity boards work. Power cuts, low voltage and poor maintenance, are problems common to the whole of India, but they vary in degree. The efficiency of the system is exceptionally low in Bihar, where there have been periods of time when tubewells would receive power for one or two hours a day, while in western UP power would be available for eight hours a day. Malpractices and defects in the power supply also seem more pronounced in eastern India.[12] 'Thefts of wire, of transformers are quite common ... like the tampering of electric installations.' In Punjab, particular effort is made to reduce the duration of power cuts.

While we cannot supply data on the extent of economic diversification in the states or large regions, the rate of urbanisation can throw at least some light on the availability of outlets for surplus rural manpower.

Percentage of urban population (1981 Census)

Punjab	27.7
Haryana	21.9
Western UP	20–25 (rough estimate)
Andhra	23.3
Tamil Nadu	33
Eastern UP	11
Bihar	12.5
West Bengal	26.5
Assam	10.3
Orissa	11.8

The rate of urbanisation in coastal Andhra should be upgraded,

[12] Reserve Bank of India, *op. cit.*, vol. 1, p. 96.

because it is higher than in the interior. As for West Bengal, it is a-typical in view of the enormous metropolitan area of Calcutta.

Development of trade, small scale industries and the commercialisation of milk are other indicators of advanced areas. One does not need data for these indicators—they strike one on driving through Punjab, Haryana and western UP. They are also quite evident through all the districts of coastal Andhra, whereas the same cannot be said of the eastern plains.[13]

These regional differences cannot be explained in economic terms alone. First is the population factor. It may have favoured Punjab, where density was and still remains lower than in many other regions in the plains. The density (1981 census) is practically the same in eastern UP (485 per sq. km.) and western UP (479). It is true that the density in the plains of Bihar and West Bengal is exceedingly high (often 600-800 per sq. km.) and usually higher than that in most advanced areas, including Andhra Pradesh and Tamil Nadu. On the other hand, where the density is more moderate as in Assam and coastal Orissa, growth has not been conspicuously fast.

While the effect of population pressure cannot be disregarded, it is not one of the major determinants of the gap between relatively rich and poor districts.

What about natural factors like the climate and soils? The climatic conditions of Punjab and coastal Andhra do not have anything in common. Nevertheless, the eastern plains face much more serious water management constraints than do the north-western plains. Excess water even in normal years, and floods during a too generous monsoon, require considerably more drainage and flood protection works here than in the north-west. If the plains of north Bihar are particularly exposed, West Bengal, Assam and coastal Orissa are hardly less prone to such dangers. The examples given in some districts are representative of very widespread conditions. 'A sound infrastructure on irrigation and drainage, which can over-come the constraints posed by physical environments, has not been developed as a whole,'[14] write the authors of

[13] Precise studies beyond the scope of this book would probably confirm these impressions which themselves fit in well with the analyses given at the district level.

[14] *Op. cit.*, vol. 1., p. 71

the Reserve Bank Report on Eastern India. The same applies to the plains of the Brahmaputra in Assam but has not been touched upon by the Report. Cyclonic damage is very severe along the coast of the Bay of Bengal, down to the coast of Tamil Nadu, hitting both sluggish and advanced districts.

As far as soils are concerned, differences are marginal, or at least much less striking. In practically all places we come across good soils, with plenty of ground water.

The historical background of each area is not without influence and would deserve a detailed analysis. Suffice here to remember that the present advanced districts were already advanced by the turn of the century, if not earlier. When starting from a higher economic level, development is obviously easier than in the more sluggish and already heavily populated regions, like most of eastern India. From this perspective the districts and villages surveyed illustrate similar trends in other parts of the north-west and the south-east. The same is true for the eastern plains in spite of certain nuances. For instance, in some districts of Bihar, south of the Ganges, there has been more progress than in the northern plains. In Bengal, some districts, Burdwan in particular, have done fairly well.

When dealing with the entire Ganges basin, one must refer to the zamindari system introduced by the British. The eastern plains were under the Permanent Settlement of Lord Cornwallis (1793), whereas from central UP to Punjab the system of Temporary Settlement was introduced in which the land revenue was reassessed every two to three decades. The former system, by giving more power to the zamindars had a negative impact on development, although such a complex issue would need a more detailed analysis.

Local power structures also enter the picture. In Bihar and eastern UP the dominant castes were — and remain — Thakurs or Bhumihars who do not have a traditionally agricultural background. In western UP, Haryana and Punjab, on the other hand the Jats played the dominant role from the 18th century onwards. They can be Sikh, Hindu or Muslim, but regardless of their religion, they are peasants *par excellence,* very hard working, industrious, ready to innovate and push intensive crops — all the qualities which had won the admiration of many British civil servants. Besides, the Jat *spirit* is not confined to their caste alone.

As seen in Bulandshahr — and this applies to many western districts — other local castes, Brahmins, Vaishyas, Lodha Rajputs, are also quite shrewd farmers.

Coming to the coastal areas along the Bay of Bengal, I cannot say much about Orissa, although one gets the impression that the high castes here are nearer in spirit to those in the lower Ganges basin than to the Jats. In Andhra, Kammas, Kapus and Reddis are clearly more enterprising. As seen in Manchala, Brahmins followed the taboo that forbade them from ploughing, but this did not prevent them from supervising their fields very carefully.

The caste factor, as an agent of agricultural growth or as a delaying factor, should not be assessed in a rigid or static way. Traditions and habits do change. Today, in many villages of eastern UP, Brahmins and other high castes use the plough themselves, whereas the taboo remains as rigid in Bihar and Orissa. On the other hand we have seen how worried Jat farmers in Mirpur are about their educated sons turning soft!

State politics is another important source of the hiatus between advanced areas and potentially rich but poor areas. Here again the matter is far from simple and one cannot merely equate political instability and slow growth. In Punjab, a considerable expansion took place under Chief Minister Pratap Singh Kairon in the late 1950s and early 1960s, but the breakthrough of the green revolution occurred at a time of great political instability, between 1967 and 1971. Later on (and particularly now), agriculture kept progressing in spite of growing disturbances. On the other hand, if the present turmoil goes on for several more years, it is finally bound to seriously affect the economy of Punjab. Especially in urban areas, the pace of industrialisation is already shaky.

Uttar Pradesh is an even more striking example. Political instability began in the early 1960s, when Pandit Pant, the then Chief Minister, left to join the central government. Since then the political scene has been in a state of flux. Yet, how can one explain the development of western UP as compared to eastern UP?

We can give some very partial and tentative answers. First, in both Punjab and western UP, the caste pattern was favourable to agricultural growth from the very beginning. Second, development could rely on private initiative, especially the construction of private tubewells, which has been such a crucial factor in the process of growth. Further east, more state intervention and planning was — and is still — needed for drainage and flood control which are in so many cases beyond the capacity of single farmers and of one village alone.

In the case of Bihar, apart from water management constraints, politics has been a major handicap to any development activity for more than two decades. Instability, feuds of all kinds involving castes and classes and the lack of any real local leaders, have created enormous obstacles to any sustained development policy. It is even more worrying to see that the deterioration process does not slacken leading to frightfully bloody incidents, especially in the plains south of the Ganges: growing clashes between, on the one side the high caste Bhumihars and Thakurs, Brahmins and Kayastha[15] and, on the other side, the rising middle castes and classes like the Kurmis and Yadavs. At the same time, incidence of violence and tension occur between high castes and Scheduled Castes, between the Scheduled Castes and the Kurmis-Yadavs. Armed gangs on all sides are becoming a common sight in certain areas. Though not of the same magnitude, such bloody incidents are not alien to eastern and central UP.

The grave disturbances in Assam have also prevented the implementation of any strong development policy, but they cannot bear the sole responsibility for the present economic situation. When I visited Assam in 1978, when the real trouble was just beginning, the economy was and had been already sluggish for many years.

Orissa has not known such deep crises, but one weak government has followed another for the past few decades or more.

Andhra and Tamil Nadu have, on the whole, enjoyed greater stability with a relatively strong political leadership—at least till the end of the 1970s. Nowadays the political hold of the ADMK is weakening in Tamil Nadu, and it remains to be seen what will be the final results of the policy followed by N.T. Rama Rao in Andhra.

Closely connected to political leadership is the administrative framework, a matter to which I will return in the next chapter. During all my surveys, from the early 1960s onwards, both in the districts and in the state capitals where I met several senior officers connected with development, I did not find significant differences

[15] This of course does not mean that are no clashes between the high castes themselves.

in the calibre of IAS officers from one state to the other. What varies more is how their abilities are put to use. It is evident that the administration works more smoothly under a rather firm and stable political leadership (than under incompetent politicians). In such cases all kinds of situations can be found: officials become victims of political interference, frequent transfers, and with corrupt practices. Not uncommon, are instances where the civil service partially makes up for the deficiencies of the politicians, ensuring a minimum of stability. It would, however, be wrong to lay all the blame at the politicians' door. One comes across opportunistic IAS officers who are more concerned about their careers rather than with public service. Others may lack tact in their dealings with politicians or object to being stationed in districts.

Other services are also affected, especially the ones which control large amounts of money. Malpractices and inefficiency are evidently connected with state politics. One instance of this is the particularly poor performance of the state electricity board in Bihar. The public works department, and other departments can also be seriously affected, like cooperative societies, cooperative banks, and commercial banks. In all these matters one comes across substantial differences between states and this has been so for many years.

The pattern of landholdings is not the same everywhere. As seen in Bulandshahr, and Varanasi districts, the consolidation of holdings has positively affected overall development and in particular, the construction of private tubewells. Punjab and Haryana had completed these operations much before Uttar Pradesh where now, most districts are covered. While Orissa has began land consolidation, many other states seem reluctant or hesitant to do so. It is true that it is a very tedious and time consuming job, opening the way to additional land litigations and malpractices of all sorts. I was told that when they made some attempts in Bihar, it was quickly discontinued because it was like a Pandora's box in terms of eruptions of litigations, and other conflicts. And yet, a consolidation of holdings would not only encourage minor private irrigation, it would also help update land records, which are very much out of date in several states.

Some officials in Tamil Nadu stressed a point which applies even more to the eastern plains: wherever you have low lands exposed to floods or at least excess water, and, other higher lands, consolidation should enable people to keep more than one plot, depending on its location, which creates additional difficulties in the consolidation operations. Or one could envisage simultaneous drainage works-cum-consolidation.

Compared to the north-west, the average size of landholdings tends to be smaller in the eastern plains, along the Bay of Bengal, down to Tamil Nadu.[16] There are also more landless labourers here, as is evident from all reports and censuses. The examples drawn from the field fit in well with the data for larger districts or even states.

This factor may have made development easier in the north-west, especially in Punjab, where the average size of holdings is larger, but the same factor has not prevented several south-eastern districts of Andhra and Tamil Nadu from faring much better than eastern India.

A much more troublesome problem than size is the effect of tenancy, sharecropping, fixed rent and other arrangements on productivity, compared to cultivating owners. Agricultural censuses and other surveys are not too reliable because of the problem of concealed tenancy, the extent of which cannot be measured. For instance, the *Agricultural Census 1971* (p. 439) mentions a total of 7.57 million operating holdings in Bihar, 7.54 million of which are wholly self-operated and owned. According to Dr. C.H. Hanumantha Rao, member of the Planning Commission and a renowned expert on India's agriculture, 'it has been found that 50 to 60 per cent of the land under rice in the eastern states is under sharecropping.'[16] How can we reconcile these enormous differences?

I would be tempted to conclude that tenancy is more prevalent in the eastern plains than in the north-west. Comparisons cannot be drawn between the eastern plains and the south-east deltas because though the villages visited in Thanjavur and Guntur had little tenancy, this may not be true for all places. One must also ascertain the type of tenancy and its way of operating. There is now a trend for landowners to take more interest in field operations by bearing part of the cost of inputs which stimulates production,

[16] One must, however, bear in mind the big landlords of Bihar who own 50-100 ha. or more something very rare, or rare in other states except Rajasthan. But in that state, many estates have poor rainfed soils with very low yields.

unlike in the old sharecropping pattern where the sharecropper had to bear all or most of the production cost for receiving only half the crop. I came across both situations during my surveys and the same has been observed by P.Bardhan and A.Rudra.[18]

One must also consider different situations that may arise. The sharecropper need not always be a small farmer. It is not uncommon to find small landowners who give out their land on a sharecropping basis to a bigger farmer, while they themselves work outside. At the all-India level, one can say that tenancy plays a minor role, the majority of the holdings being operated by cultivating owners. But, it is probably less than 90 per cent as estimated in the Agricultural Census.[19]

Peninsular India

Leaving aside soil variations and some prosperous districts, peninsular India faces the problem of soil constraints much more than the alluvial plains. Light red soils and heavy black soils are less fertile. Soil erosion caused by water and — in desert areas, wind erosion — is severe. Soil degradation encouraged by the reclamation of marginal land, deforestation and over-grazing, is a matter of growing concern, accentuated by population pressure on land.

Irrigation facilities too are limited and its potential for the future is no brighter, unlike in most alluvial plains. The conditions prevailing in the two districts surveyed (Deoghar and Keonjhar) illustrate conditions that are widespread in the north-east areas of the peninsula. The districts of the Bihar plateaus have no more than 10 per cent of their gross sown area under irrigation. In the interiors of Orissa, one comes across the same percentage for the northern districts, it rises to 20 per cent in the central part and falls to 7 per cent in the southern part. Average yields of clean rice vary between 700 and 1,000 kg./ha. (1,000 - 1,500 kg./ha. of paddy) without a clear upward trend over the past ten to fifteen years.[20]

[18] See their article, Terms and Conditions of Share Cropping Contracts, *Journal of Development Studies*, April 1980.

[19] Government of India, *Agriculture Census 1970-71* (New Delhi, Ministry of Agriculture).

[20] Reserve Bank of India, *op. cit.*, vol. 2.

Coming to the interiors of Andhra, in the Rayalaseema region where Kurnool district is located, around 15 - 25 per cent of the gross sown area is irrigated. Rice is the main beneficiary, whereas the larger area devoted to coarse grain remains exposed to erratic rains.

The limited area devoted to irrigated crops is also evident in all the states entirely or mainly located in the interiors of the peninsula.

Percentage of irrigated area to cropped area (1980-81)

Gujarat[20]	21.8
Karnataka	15.7
Madhya Pradesh	11.5
Maharashtra	12.4
Rajasthan	21.6

Source: Government of India, *Area and Production of Principal Groups in India,* (Ministry of Agriculture, 1981-84).

Here again statistics require qualification. First, rainfall varies within this huge area, so that the impact of irrigation on overall production is no less diverse. Second, the purpose of irrigation or the crops covered is not the same. In Rajasthan, Gujarat and Madhya Pradesh, the pattern does not differ much from what has been observed in the north-west: wheat and possibly some cash crops including vegetables grown near urban markets are the main beneficiaries.

In Karnataka, rice is irrigated, as also vineyards, banana, coconut trees and sugarcane.

In Maharashtra, a large part of canal and well irrigated areas are devoted to sugarcane; other plots, as seen in Satara — or observed earlier in Ahmednagar — are devoted to orchards, vegetables and chilli. The same pattern of kitchen gardens has also been noted in the plateau areas of Bihar and Orissa, with the help of ordinary wells.

Irrigation of cash crops, especially fruit and vegetables, though limited in size, can bring a better return than larger irrigated areas devoted to cereals. In a way, this can reduce the gap in irrigated areas between peninsular districts and those in the north-west or in the south-east. On the other hand, the gap may get wider

[21] Only parts of Gujarat belong to peninsular India.

because irrigation is often vulnerable to drought when the ground water supply for wells is not replenished or when the supply to canals is limited. This, of course, compares very unfavourably with a relatively safe perennial irrigation.

The consumption of chemical fertilisers is low, because it is devoted mostly to irrigated crops. We have seen how farmers cannot take the risk of using chemical fertilisers in rainfed fields in view of possible drought or insufficient rainfall.

Consumption of chemical fertilisers (nutrients) kg./ha. gross cropped area – 1981

Karnataka	38.3
Gujarat	38.7
Maharashtra	26.3
Madhya Pradesh	11.0
Rajasthan	9.1

Source: *Uttar Pradesh, Draft Seventh Five Year Plan,* p. 14.

Since then, there have been no major changes, particularly in Madhya Pradesh and Rajasthan, 'which have been lagging far behind the national average' of 50 kg./gross ha.[22]

In view of the limited role played thus far by irrigation, new seeds of cereals, chemical fertilisers, rainfall and soil are of decisive importance, especially now that there is hardly any land left for reclamation.

The Indian Meteorological Department has published rainfall normals 'based on available data during 1901 to 1950.' In its monumental report, the National Commission for Agriculture expressed the view that these data should be revised, an opinion I fully share from observations made in districts where often 'normal' rainfall occurs rather rarely![23]

In spite of possible limitations, rainfall series 'have shown no significant trend on the whole' so that one cannot talk of increasing or declining rainfall in the long term. However, 'weather as a single factor can be responsible for as much as 50 per cent variations in yield which occur from year to year, the remaining 50 per cent being due to other factors like irrigation, manuring and plant

[22] *Economic survey 1986-87*, p. 11

[23] See data for Keonjhar in chapter 10.

protection.'[24] These broad observations correspond to what we have seen in rainfed areas.

Rainfall in wide tracts of the Deccan is between 500 and 750 mm., falling to 300 mm. and less in western Rajasthan, rising to above 1,000 mm. in the north-eastern corner of the peninsula. About 80 per cent of the rainfall occurs during the monsoon.

One-tenth (317,000 sq. km.) of the total area in India lies in arid zones, two-thirds in Rajasthan, one-fifth in Gujarat, the rest in parts of Haryana and Punjab, in Maharashtra, Karnataka and Andhra. One finds in these areas conditions which remind us of the Sahel in Africa. The eco-system has been deteriorating in the past few decades, and, as seen in the field, remedies seem costly and complex.

Other parts of the peninsula are also suffering from soil deterioration through erosion, increase in waste land, reduction of pasture land or over-grazing and deforestation. A number of schemes are under design or implementation, but the task is of such magnitude that it will take decades to substantially stop the damage and improve soils. All these weaknesses are associated with rainfed crops because fields under irrigation are bunded to retain the water. A total cultivated area of 87 million hectares are exposed to soil erosion, to which one should add desert and coastal areas exposed to wind erosion.[25]

Jowar, bajra, pulses and rainfed paddy are by far the major crops in the interior of the peninsula:

Main foodgrains sown areas (million ha.)

	Peninsular India	All-India
Jowar	14	16
Bajra	10	12
Maize	2	6
Rice*	10*	41
Pulses	14	23
	50	98

*mostly rainfed.

Source: *Agricultural Census 1970-71 op.cit.*, Area and Production of *Principal Crops in India 1981-84, op.cit.*

[24] Government of India, *Abridged Report* (New Delhi, National Commission for Agriculture, 1977), pp. 153-59.

[25] *Ibid.*, p. 218.

The major rice growing areas lie in the plateaus of southern Bihar and Orissa (600 – 1,000 kg./ha. of clean rice, 900 – 1,500 kg./ha. of paddy) as well as in Madhya Pradesh (1,000 – 1,500 kg./ha., average paddy including 18 per cent of the area irrigated).

One-third of the total *jowar* area is found in Maharashtra where yields were around 350-400 kg./ha. in the early 1960s. Now, in a good year the yield is around 700 kg./ha. In Karnataka and Madhya Pradesh it now fluctuates between 700 and 900 kg./ha., in Andhra between 560 and 720 kg./ha.

For *bajra*, yields remain more erratic. Forty per cent of the sown area is in Rajasthan. Under good rains (1983-84), the state average yield reached 500 kg./ha., as against 290 kg./ha. the year before and probably less in 1985-86. In 1982-83 and 1983-84 it was 320 kg./ha. and 500 kg./ha. in Maharashtra, 600 kg./ha. and 700 kg./ha. in Madhya Pradesh and 370 kg./ha. and 500 kg./ha. in Karnataka.

All these data show some increase in yields but this increase is modest, when bearing in mind that *jowar* was at 450-530 kg./ha., and *bajra* at 250-330 kg./ha. in the mid-1950s.[26]

Although pulses are grown in the alluvial plains, they are found more frequently in peninsular areas and are rainfed. Madhya Pradesh ranks first with 5 million ha. followed by Maharashtra with nearly 3 million ha. The average state yields for 1983-84 (a good year) were 530 and 425 kg./ha. respectively. There has been no noticeable increase in yields over the past fifteen years. The all-India data oscillate between 380 and 550 kg./ha.

Wheat is relatively important in Rajasthan and Gujarat, where 75 to 80 per cent of the area is irrigated, producing 2,100-2,400 kg./ha., whereas in Madhya Pradesh it falls to an average of 1,200-1,600 kg./ha., including both irrigated wheat (30 per cent of the area) and non-irrigated wheat.

To return to the all-India picture, about 70 per cent of the total cultivated area is rainfed, the bulk of which is in the Deccan and supplies 40 per cent of the total foodgrain production.[27]

Oil seeds (on 18 to 20 million ha. according to the weather, out of which 15.7 to 16.6 million ha. occupy edible oil seeds), are more widespread in the Deccan than in the plains. Approximately

[26] See also the data given earlier for all India production.

[27] Government of India, *Seventh Five Year Plan, op. cit.*, p. 9.

10 per cent of the fields are irrigated. There has been some increase, especially in the last few years. New crops like soyabean have been introduced, and have been successful in Madhya Pradesh and could even expand further. However, the lack of irrigation prevents sizable yield increases and the all-India deficit in oil seeds implies a heavy drain of foreign exchange.

Cotton is another important crop in peninsular India. Out of 7.4-7.9 million ha., 2.7 million ha. are to be found in Maharashtra with a very low yield of 70-100 kg./ha. of cotton lint. Gujarat follows with 1.4-1.5 million ha.(175-230 kg./ha.) and Karnataka with 0.9 million ha. and 100 kg./ha. Once again comes the problem of lack of irrigation: in Maharashtra 5 per cent of the cotton fields are irrigated; 7 per cent in Karnataka; 28 per cent in Gujarat. In the past twelve years output at the all-India level has increased by about 40 per cent only.

What is the situation of the other sectors of the economy and how do they compare with the plains? As regards infrastructure (roads and rural electrification), the situation is as follows:

	Surfaced road length per 100 sq. km.	*Percentage of electrified villages*
Gujarat	21	76.8
Karnataka	33	75.7
Maharashtra	26	85.5
Madhya Pradesh	12	46.3
Rajasthan	10	52

Compared to more advanced regions, the first three states do not fall behind in rural electrification, but the same does not hold for the roads, except in Karnataka. The other two states are nearer the states in eastern India.

With regard to the global process of economic diversification, the role played by urbanisation and industrialisation as observed in Satara is also felt in several other districts of Maharashtra. It is quite clear in Karnataka and Gujarat, where industry is important. In the latter, small industry is expanding rapidly in rural areas. In 1980-81, Maharashtra accounted for 24 per cent of the total industrial output in the country, Gujarat for 10.6 and Karnataka for 4 per cent.

Both Madhya Pradesh and Rajasthan have fewer outlets for their agricultural population. In this respect, the case of Jodhpur district is not really representative because of its very high rate of urbanisation as compared to the state average.

Urban population as percentage of total (1981)	
Maharashtra	35.0
Gujarat	31.1
Karnataka	28.9
Madhya Pradesh	20.3
Rajasthan	21.0
Plateau areas of Bihar and Orissa	10 - 20

There is some homogeneity in density per sq. km., between Karnataka, Maharashtra and Gujarat (approximately 200 per sq. km.), whereas it falls to 120 per sq. km. in Madhya Pradesh and 100 per sq. km. in Rajasthan. Maharashtra and Gujarat have larger tracts of semi-arid or drought prone areas than does Karnataka. Gujarat, and more so Rajasthan, also have wide desert or semi-desert areas. Other parts of the Deccan, however, enjoy better rainfall, especially the eastern part of Madhya Pradesh and Vidarbha in Maharashtra. As a result, the size of holdings is usually larger than in the plains but with such low productivity except in irrigated plots, that people are not able to live very well.

Not much progress has been made towards land consolidation, as it may be less beneficial here than in areas endowed with a high irrigation potential. It will also be necessary to allow farmers at least a few plots of red and black soil both, wherever they may be within the same village (see Karwad).

Forest areas must also be considered. They cover wide tracts in the north-eastern part of the peninsula. They do bring some additional income (from wood, wild fruit, *tassar*, grazing) to the villagers who live near by. On the other hand, forest resources are dwindling in spite of recent efforts at afforestation.[28] Further, forest areas are often inhabited by *adivasis*, many of whom still practice shifting and burning cultivation which leads to deforestation as a result of population pressure.[29]

[28] Recent studies have shown that real forests have fallen from 55 million ha. in 1975 to 46 million ha. in 1982. See A. Ghose, 'The Second Green Revolution,' *Courrier de l' Inde*, 15 May 1986.

[29] I will return to this problem in the following chapter.

What is the role of castes? A number of them, the Reddis in Andhra, Marathas in Maharashtra and Patidars in Gujarat and parts of Rajasthan, enjoy a long-standing and firm tradition of agriculture as a result of which one does not come across those cultural factors which may inhibit agriculture as in the eastern plains. But, I cannot say more about many parts of the Deccan for lack of experience, especially about Karnataka and Madhya Pradesh.

For many years Karnataka, Maharashtra and Gujarat enjoyed a much more stable and able political leadership than did UP or Bihar. As a result, administration was also more efficient. No doubt one could come across malpractices, wastage or local tensions, but again, on a rather low key. As far back as 1978-79, some IAS officers confessed that the difference between working in the above states and in the northern states, if not Orissa, was comparable to the difference between day and night.

In both Gujarat and Maharashtra the political scene has been gradually deteriorating since the 1980s, if not a little earlier. There have been frequent changes of the chief minister, factions, feuds and, grave communalist tensions erupting into violence. From what I could see in Maharashtra in early 1986, one nevertheless gets the feeling that the administration can still partly mitigate the erosion of political power. Besides, the political scene has improved with the advent of S.B.Chavan as Chief Minister.

Karnataka has managed to fare better till now, first under Congress governments and now under the Janata Party.

Madhya Pradesh is a particularly difficult state to run, due to its size (the largest state in India), and its background: a large part of it was composed of princely states and another part belonged to the central provinces in the days of British rule. Rajasthan too was entirely carved out of princely states after Independence. On the whole, Rajasthan has done reasonably well, perhaps better than Madhya Pradesh, especially when considering the amalgam of many feudal states, some rather advanced like Jaipur, others under not very enlightened rulers!

For the period since Independence, one would be tempted to conclude that the political climate has not really hindered development in the Deccan so far, but the deterioration process, if not checked in Maharashtra and Gujarat, can, in the long run, affect rural development.

The impact of growth on poverty

The link between overall growth and diversification of the local economy on the one hand, and conditions of living on the other, is clearly evident from our field studies. Higher yields per hectare and multi-cropping go along with higher wages. In progressive areas, daily wages are now frequently around Rs. 10-12 (1985-86), sometimes more, or amount to 10-15 kg. of grain per day. In the more slow moving areas, either in the eastern plains or in peninsular regions, wages fall to Rs. 5-7 or 3-5 kg. of grain per day, if not less. While it is impossible to measure employment opportunities precisely, it can be said that there are more outlets if the process of economic diversification is extensive enough.

Several more general studies, which had already appeared during the second round of my surveys in 1978-79, confirmed these trends. Unfortunately, the Indian scholars referred to[30] did not enjoy the same publicity as Francine Frankel who claimed that 'landless labourers ... have been excluded almost entirely from participating in the benefits of the green revolution', although she gives instances of rising wages in Ludhiana district in Punjab![31]

Later studies also confirm our own findings. M.L. Dantwala underlines the diversification of the Punjab economy and its social impact where 'the share of workers in the secondary and tertiary sectors has increased from 37.3 per cent in 1971 to 40.8 per cent in 1981.' He mentions also how even small farmers are resorting to new inputs. C.H. Hanumantha Rao, D.B. Gupta and P.S. Sharma suggest that: 'The development of irrigation seems to be associated with a lower poverty ratio.' Montek S. Ahluwalia, on studying the National Sample Survey data, comes to the conclusion that there has been 'a more or less steady decline in poverty' after the green revolution, but there are, of course, differences among states.[32]

[30] B. Sen, *The Green Revolution in India* (New York, Wiley, 1974). C. H. Hanumantha Rao, *Technological Change and Distribution of Gains in Indian Agriculture* (New Delhi, 1975). C. H. Shah and C. N. Vakil eds., *Agricultural Development of India*, (Bombay, Orient Longman, 1979).

[31] F. Frankel. *India's Green Revolution*, (Princeton University Press, 1971), pp. 105 and 37.

[32] For all these quotations see the remarkable book by J. W. Mellor and G. M. Desai eds. *Agricultural Change and Rural Poverty: Variations on a Theme by Dharm Narain* (Delhi, Oxford University Press, 1985). For job diversification see also V. S. D'Souza, *Economic Development, Social Structure and Population Growth*, (New Delhi, Sage 1985).

When I discussed these issues with Pradhan Prasad at the A.N. Sinha Institute for Social Studies in Patna, he also agreed that there had been some progress in real wages for the poor in advanced districts. B.M. Bhatia writes that 'The states with [a] higher growth rate of agriculture have fared better than others in improving the living conditions of the people in the last three decades'.[33] In his recent book, G.K. Chadha highlights, among other issues, the benefits accruing from the green revolution to the small farmers and landless labourers in Punjab as well as a large number of workers who have come to Punjab from UP and Bihar.[34]

I could also quote a large number of experienced senior officers in New Delhi and other state capitals and districts who also endorse the views expressed here.

Although it has taken a long time, the sad and so often quoted slogan of the green revolution, 'the rich get richer and the poor get poorer,' is being at last seriously eroded in India's public opinion thanks to a better awareness of how the poor do benefit from it, provided the economy diversifies fast enough. As we have seen, their progress is not spectacular, but their future is no more closed or blocked. Gradual improvements have been made in their living conditions, whereas in sluggish regions, even limited progress for the poor is not assured.

To summarise the achievements in the forty years since Independence, one could say that the battle against poverty is half won. The danger of mass starvation deaths has been checked for the first time in India's history. Technical innovations are spreading even to the remote parts of the country, but the pace at which they are adopted varies enormously. Acute poverty is retreating, at least in the advanced districts, but even in the more slack areas, one comes across at least some poor people who are now less insecure than before.

On the basis of what has been achieved, what are the prospects for the future? To try and answer that question, we must consider the main aspects of the development policies followed so far and what changes are being — or could be — introduced.

[33] B. M. Bhatia, *Indian Agriculture, A Fifteen Year (1985-2000) Policy Perspective,* (New Delhi, Centre for Policy Research, October 1985).

[34] G.K. Chadha, *The State and Rural Economic Transformation–The Case of Punjab,* (New Delhi, Sage, 1986).

Chapter Fourteen

Rural Development Policies and the Attack on Poverty

Is it correct to claim, as is often done, that rural development and poverty have been neglected for too long? In the 1970s, a whole body of experts and scholars 'discovered' poverty when criticising the 'trickle down effect' where growth, according to them, had not alleviated poverty as expected. A group of Asian scholars wrote about 'a quarter century of (anti) rural development'. Others criticised the green revolution' and the use of modern technology.[1]

As Jagdish Bhagwati has pointed out, there have no doubt been mistakes or errors when 'discussing the link between growth and poverty' in the early phases of planning in India as in other developing countries. 'But it is no more than vulgar nonsense to allege that, therefore, growth was regarded universally in itself and that poverty was not in the picture.'[2] In fact, the report of the First Five-Year Plan of India (1951-56) had already conceived rural development in a broad economic *and* social framework.

The main currents of rural development policies

The concern for rural development and for the eradication of poverty were already playing a prominent role in the Congress movement, even before the advent of Mahatma Gandhi. In the 1920s and 1930s, many political leaders, and Nehru the first of them, were 'ruralised,' touring villages, discovering a world quite different from the one which the urban intelligentsia was accustomed to.

[1] We cannot quote the numerous publications on these topics. Among others, see W. Haque, N. Mehta, Anisur Rahman and P. Wignaraja, 'Towards a Theory of Rural Development, *Development Dialogue,* no. 2, 1977. See also *Poverty and Landlessness in Asia,* an I.L.O. Study, Geneva, 1977.

[2] J.N. Bhagwati, in G. Grossman ed., *Wealth and Poverty* (Oxford, Basil Blackwell, 1985), p. 17

Following the 1937 elections, the provincial governments showed a keen interest in rural development. Many surveys were conducted and programmes started in order to accelerate and widen the efforts made so far in a too limited way.

This concern found new and more powerful means after Independence. The Indian leadership envisaged a three-pronged attack : land reform, particularly the abolition of the zamindari system; the creation or expansion of rural institutions — community development projects and extension, credit cooperative societies; and, at the end of the 1950s, Panchayati Raj, i.e., the creation of rural councils at village, block and district levels. Irrigation, development of rural infrastructure (roads, electricity) land reclamation and modernisation of agriculture were no less encouraged.

From the 1960s onwards, rural development entered a new phase with more emphasis on technical innovations, new seeds, chemical fertilisers and pesticides. As for irrigation, more efforts were made in the field of minor irrigation (tubewells, pumpsets) while major projects were carried further. At the same time, more investments were concentrated on the most promising areas (alluvial irrigated plains) from which quick results could be expected in terms of production.

If one cannot say that the poor were disregarded in the 1960s, it is correct to say that greater emphasis on social justice re-emerged in the 1970s around the politically fraught slogan of '*garibi hatao*' (to remove poverty). To the existing policies, special agencies and schemes for the poor were added the Small Farmers Development Agency (SFDA), the Drought Prone Areas Programme (DPAP) and various food for work programmes as well as the Command Area Development Agency (CADA) designed to speed up the development of newly irrigated areas under major projects. The basic idea was to combine a direct approach to poverty or the 'target group approach' and the indirect approach through overall economic growth. One should not, however, forget the political aspects of such schemes, as obvious now as they were in the 1970s. They make room for various political manoeuvres and local patronage.

At the planning level, the early debate on priorities either to agriculture or industry became obsolete in the 1960s and remains so today. By that time it gradually became clearer that the key issue was not priorities but the best combination of pull and push effects between agriculture, industry and infrastructure which were becoming increasingly interdependent. Agriculture needed — and even more so now — a strong support from industries supplying a number of inputs : fertilisers, pesticides, implements, pumps and motors, tractors, cement for water works, steel, electricity, roads etc.

At the same time, several industries like textile mills, sugar mills, food and leather industries, needed a larger supply of agricultural goods.

All these efforts have produced the results presented in this book, but they need some qualifications. The new rural institutions (cooperatives, community development, Panchayati Raj) presented, even at the time of my first survey,[3] a mixed picture which has not fundamentally changed today. The first two did contribute to rural expansion, but they remained below the original and not very realistic expectations : malpractices, lack of support to the poor because of the weight of medium and large farmers in these institutions, were some of the shortcomings. The record of Panchayati Raj has not been very encouraging so far. In several states (particularly Bihar and UP), they have not been very active due to infighting, local feuds and rivalries. Where they did work — in Maharashtra, Gujarat and Tamil Nadu — they were dominated by large and medium farmers as a result of which the poor did not derive much benefit.[4]

The agricultural policy proper did achieve substantial results: there has been progress in research and seeds multiplication, growing improvements in the supply of inputs like fertilisers. In this regard, one still comes across weaknesses, especially in the eastern plains, but today one hears much fewer complaints from farmers in the advanced areas regarding delays, shortages leading to malpractices, etc.

The increase in irrigated land is no less impressive, again, in spite of various shortcomings, a matter to which I will return later, especially with regard to major works.

[3] See part 2 of my book, ***Studies in Indian Agriculture*** (Berkeley, University of California Press, 1968).

[4] See among other reports, Government of India, *The Mehta Report* (New Delhi, Report of the Committee on Panchayati Raj Institutions, 1978)

All in all, from the district to the state capitals and up to New Delhi, there is, today, much more knowledge and expertise about rural development and about technical, social and economic issues than there was twenty or thirty years ago. This shows that through a process of trial and error, a lot has been learnt and has helped in correcting mistakes, while improving the whole development process.

Repeated land reforms

It is surprising to see the frequent re-emergence of the question of land reforms when, even twenty years ago, it looked like a blind alley for political and technical reasons. As long as State Assemblies are dominated by a large number of, not always big landlords, but at least large farmers, there is no scope for drastically lowering the present ceiling on holdings which had been adopted in the 1970s : from 6.3 ha. of good irrigated land to 10.8 ha. with only one irrigated crop per year and no more than 21.6 ha. of other land. It was estimated at that time that 2.35 million ha. of surplus land would be made available. This small area was totally out of proportion with the land needed for all the landless labourers and tiny landholders below 0.5 ha. (see previous chapter). As a left-leaning Indian economist told me in 1978, through the process of division of holdings, big landlords and many large farmers are fading away, a point illustrated by the agricultural Censuses of 1971 and 1981.

Then, the implementation of ceiling laws meets with considerable obstacles, such as the division of holdings 'on paper' by several members of a family and all sorts of litigations. The actual declared surplus was only 1.74 million ha., out of which 902,000 ha. had been distributed by 1985. Nearly 655,000 ha. are 'locked in litigation.'[5]

The prospects do not appear any better with reference to tenancy laws and minimum wages because they will be very difficult to implement. Sharecroppers and landless labourers would not dare to complain if the laws are not respected for fear of losing their sharecropped land or their job.

[5] Government of India, *New Vista in Land Reforms* (New Delhi, State Revenue Minister's Conference, 18 May, 1985, Ministry of Agriculture).

One must, however, make two exceptions. In Kerala and West Bengal, as a consequence of a strong communist party often in power for many years (as in Calcutta), land reforms were pursued and implemented more strictly than in most other states.[6]

Other reforms look more promising because they face fewer political obstacles. One such is the release of homesteads for the landless. This is already being done in most states and, as observed in the field, though a modest step, it is not negligible for the concerned person who feels less dependent and less open to exploitation than when he was living on another's land, often his employer's.

A second step would be the thorough updating of land records, which are in a particularly bad shape in areas formerly under the Permanent Settlement of eastern India, which often had adverse effects on small farmers. It looks like a very tough job, as seen in Bihar. Besides, in certain states like Andhra Pradesh, the policy is now to abolish the post of *patwari*, village accountant in charge of land records, and replace it by officials who have no experience in that delicate job.[7]

The question of consolidation of holdings could be added to land records updating. We have seen its impact on private irrigation in Punjab, Haryana and UP. On the other hand, some states are reluctant, especially in areas where parts of the fields are exposed to excess water or floods. There might also be problems in villages of peninsular India where often one comes across a variety of soil conditions. In 1985, the consolidated area was 52 million ha. (about 33 per cent of cropped area) according to the Seventh Plan Report (p. 63).

The situation of rural institutions and agencies today

During the Sixth Plan (1980-85), the Command Area Development Agency (CADA) was retained while the new agencies were realigned. The pattern of extension within the community

[6] On land reforms see R.J. Herring, *Land to the Tiller: The Political Economy of Agrarian Reforms in South Asia* (New Haven, Yale University Press, 1983).

[7] The plea for updating land records comes up frequently, without much being done. Raised in the 1970s if not earlier, it came up again at the Conference of Revenue Ministers in New Delhi in May 1985.

development block was altered in several states which, under World Bank guidance, adopted the Training and Visit System.

The development of these new institutions explains why 'in terms of instrument of planning, community development and Panchayati Raj institutions have been reduced to an extremely peripheral status with the budget provision in many cases being limited to maintenance of staff.' (Seventh Plan p. 64)

The present situation can be summarised as follows : the cooperative movement remains very important with 94,100 primary agricultural credit societies and 63.5 million members, 36 per cent of them being borrowers.

1984-85	*billion Rs.*
Short term loans	25
Medium term loans	2.5
Long term loans	5.0

It is distressing to read in the Seventh Plan that 'there is no perceptible improvement in the recovery of loans, and overdues have continued to be as high as 41 per cent of the demand' at the primary societies level. For land development banks (long term credit) the percentage of overdues is 43 per cent (1982-83). As a result, many societies and cooperative banks become crippled because apex institutions stop providing them with funds.

Primary societies (i.e., at the village level) are not confined to credit, since they supply inputs : 47 per cent of chemical fertilisers are distributed by them. They are also engaged in marketing operations in the order of 27 billion rupees in 1984-85. In both fields, defects and malpractices are not uncommon as it is in loan operations.

What is disturbing is to see that the same weaknesses have come up repeatedly, in all the plans since the 1960s.[8] Again, in its 1985-86 Report, the Department of Agriculture complained of 'mounting overdues' (p. 126).

Since their nationalisation in 1969, commercial banks have become more active in villages than they were before. It is now quite common to see, as in Unchagaon and other big villages or small towns, a recently opened branch of some bank engaged in loans to farmers or in the IRDP loan-cum-subsidy system. From

[8] See references in my book, *Studies in Indian Agriculture, op. cit.*

9,000 (1970), the number of banks and branches has risen to 50,000 in 1985. In 1984-85, short term loans amounted to Rs. 11.1 billion, medium and long term loans to 14.5 billion, a total increase of 150 per cent over 1980-81.[9] Yet, the quantitative boom goes along with a deterioration in management and operations. Political patronage, various types of malpractices as observed in the field, raise frequent criticism, even in official circles. More serious are the rising overdues which prevent banks from proceeding with new loans.

The old extension services under the community projects pattern have been replaced by the Training and Visit System. The district joint director of agriculture is in charge along with a number of field specialists (soils, fertilisers, pesticides, etc), agricultural extension officers and agricultural village workers. The single line of command is appropriate. It relies on frequent visits by each extension worker to a number of contact farmers in order to spread technical innovations.

The reservations I have expressed in 1978-79[10] about the system have been confirmed, at least in a number of states where the new system has not produced particularly good results. The Training and Visit System per se does not guarantee success. Like any extension system it depends above all on the calibre of the concerned officials. When one finds a splendid agricultural officer as we did in Puri (see chapter 9), the system does help, but I wonder whether similar results would not have been achieved by the same man under different arrangements, a point that my friend in Puri would not agree with! In many places, field specialists are either lacking or inefficient. As far as the regular visits to contact farmers under a fixed schedule armed with a notebook is concerned, it remains to be seen how diligently it is done. The familiar scene of the old system where the village level worker would sit and drink tea with a local *neta* (village leader) has not been completely replaced by the Training and Visit System.

While not denying some possible improvement, I gathered the same impression as I did in 1978-79: that the system per se is

[9] Government of India, *Annual Report, 1985-86* (Department of Agriculture and Cooperation, Ministry of Agriculture) p. 125.

[10] Gilbert Etienne, *India's Changing Rural Scene* (Delhi, Oxford University Press, 1982), part 2.

overstated by the World Bank, which, with great zeal, is spreading it in several developing countries under the guidance of the Israeli expert, Daniel Benor.[11]

Let us come to the target group approach and the various schemes for the poor. The main one is the Integrated Rural Development Programme (IRDP) followed by two schemes offering employment in various rural works to poor people during lean periods of the year : these are National Rural Employment Programme (NREP) and Rural Landless Employment Guarantee Programme (RLEGP). Under both programmes part of the wages are paid in kind. The Training of Rural Youth for Self-Employment (TRYSEM) scheme provides poor people with skill training in various trades. In concerned areas are the Drought Prone Areas Programme (DPAP) to improve dry land farming and the Desert Development Programme (DDP) for desert regions as seen in Jodhpur.

All these organisations are covered at the district level by the District Rural Development Agency (DRDA) which, as a registered society, enjoys a fairly large degree of autonomy. Civil servants conduct the operation under a governing body which includes local MPs and MLAs. When and where they work, the Panchayati Raj institutions are involved in the programmes 'both in terms of identifying beneficiaries and monitoring the implementation.'[12] The performance of the IRDP under the Sixth Plan showed:

Number of families assisted	:	16.5 million, out of which 6.5 million belonged to Scheduled Castes
Budgetary expenditures	:	16.6 billion rupees
Credit mobilised	:	31 billion rupees

The first item concerns administrative costs and subsidy.

[11] D. Benor and M. Baxter, *Training and Visit Extension* (Washington DC, World Bank, 1984) A. Ghosh reaches rather similar conclusions to mine on Training and Visit in his article 'Agricultural Research and Rural Development', *Economic and Political Weekly*, 4 April, 1987.

[12] See on all these schemes D. Bandhyopadhyay, *For the Rural Poor, Gramin Vikas* (New Delhi, Ministry of Agriculture, 1985). A very well presented booklet prepared by D. Bandhyopadhyay, Secretary Rural Development, New Delhi, August 1985.

To qualify, a family must earn no more than Rs. 3,500 per year, and in the case of a landowner, to own no more than 2 ha. The subsidy component varies from 25 per cent to 50 per cent of the total money allotted, depending on the situation of the person. Credit is made available on a priority basis by commercial and cooperative banks at a subsidised rate of 10 per cent (against the normal of 14 per cent) or sometimes even less.

Several reports confirm what has been observed in the field. The scheme has clearly been successful for people who, without special assistance, could not have broken the wall of misery surrounding them. However, several defects are presented in the Seventh Plan Report (p. 53-55) which confirm the other side of the story. The extent of wrong identification is set at 15-20 per cent of the assisted people. 'Leakage through corruption and malpractices' cannot be quantified but complaints are frequent and, as seen in the field, it can happen at various levels of the application procedure, a *baksheesh* to the veterinary, to the bank, to the block development officer, badly conceived assistance such as insufficient investment, or investments which the beneficiary cannot sustain. The Seventh Plan corroborates others critical of an 'overwhelming bias toward animal husbandry,' while overlooking the not infrequent lack of fodder, or the shortage of animals of good quality.[13] For other schemes, such as starting a small business, 'the absence of backward and forward linkages' is regretable. Finally, in certain areas, like the north-east, the banking infrastructure is too weak to cope with the number of applications.

The Seventh Plan Report, after reviewing these shortcomings, makes a plea for a better selection of beneficiaries with a household income not exceeding Rs. 4,800. While reducing animal husbandry projects, it is intended to 'step up activity in the industries, services and business' (p. 55), a proposal which, as seen in Jodhpur, will not be so easy to implement.

The Seventh Plan aims at providing assistance to 20 million households. The expenditure for subsidies, administration and infrastructure should amount to Rs. 26 billion for the IRDP and allied programmes, half of it from the central government, half from the states. The loan component should be around Rs. 40 billion (p. 56-57).

[13] A criticism that I heard in most states.

The Seventh Plan (p. 64) intends to encourage decentralised planning further by reactivating or strengthening Panchayati Raj institutions. In some states (Tamil Nadu and Karnataka) the system has already been reactivated through elections which had been due for many years. It is true that over the past few years one has come across instances where the system has shown improvements. However, one is a bit hesitant to place too high an expectation on these councils. On the one hand, a greater awareness of community interest may emerge. On the other, class and caste conflicts do not seem to be on the decline, as they should regardless of the diversity of this country.

The two employment schemes also present a mixed picture of defects, leakages and achievements, and it is not easy to take the number of mandays created at their face value: 351 million for the last year of the Sixth Plan. On the other hand, from what I saw in Kutch and in Rajasthan, I got the impression that the rural works started under condition of emergency during the drought may have been more successful than for more routine operations.

I must confess that I could not really study these rural works in detail but I wonder, in a very tentative way, whether they bring much hope, *except under conditions of emergency*. These formulae have been tried since the beginning of the five-year plans, always with limited success.

One thing, at least, comes out clearly: the growing importance of schemes focusing on the poor.

Planning for rural development as a whole

It is significant to see how, in the light of experience, the overall development policy has improved, with considerable effort in spreading technical innovations and in infrastructure. Progress in irrigation has been considerable, like the supply of inputs.

On the basis of what has been achieved, how does the future look?

	1984-85 Base level	*Seventh plan target*
Foodgrain output	150	178-183 million t.
Oilseeds	13	18
Sugarcane	180	217
Cotton ginned	1.27	1.67
Jute	1.35	1.71

	Actual irrigated area	Potential created (gross)
1984-85	60.4 million ha.	67.9 million ha.
1989-90	71.3	80.8

	Actual fertiliser consumption million t. (nutrients)	Pesticides material in 'ooo t.
1984-85	8.37	50
1989-90	13.50 - 14.00	75

Source: *Seventh Five Year Plan* Vol II, p. 5, 72, 81.

In spite of growing efforts for rainfed crops, irrigation combined with new seeds, chemical fertilisers and pesticides is bound to play a decisive role. The maximum progress is expected from irrigated crops while for coarse grains — mostly rainfed — the Plan expects an increase of only 2-3 million t. in order to reach 34-35 million t.

The overall target for foodgrains belongs to the art of the possible, in view of future progress in irrigated wheat and rice. The same can be said of sugarcane, mostly irrigated. The outlook for cotton, in view of the large area under rainfed conditions, seems more precarious. The same is true of oilseeds and pulses which are also bound to remain mostly rainfed.

Water management: a crucial issue

In the field of irrigation, the Seventh Plan focuses on the completion of unfinished irrigation projects and restricts new medium projects in drought prone areas, tribal and backward areas (p. 71). The construction of field channels, land levelling and the replacement of field to field irrigation by direct supply of water to the fields on a turn basis *(warabandi)* as practised in the north-west is encouraged. Maintenance of canals is to be improved through 'adequate financial allocations.' Further efforts are contemplated for the exploration and exploitation of ground water.

All these steps are in the right direction. During discussions in state capitals and in New Delhi, one got the impression that there is a greater effort than before to correct the long standing defect of fund allocation, whereby new projects keep on absorbing large amounts of money at the cost of maintenance, repair, realignment, improvements of the existing—both old and more recent—canal systems. One must bear in mind that investment budgets are devoted to new projects, while the ordinary budget deals with maintenance and operations. Until now, in case of budgetary constraints, a frequent situation, the ordinary budget was subjected to more cuts than the investment budgets. In quite a few states and in New Delhi, officers expressed their concern on the fundamental problem of plan and non-plan expenditure, which is unfortunately not confined to irrigation alone.

I wonder whether the changes proposed in the Seventh Plan are radical enough for two reasons. The defects in fund allocation between investment and other expenditures have been identified for at least the past fifteen years, which means that *drastic changes* would be needed, a task which is not easy. Too often, rational decisions are prevented or altered under the pressure of short-sighted political interests, going along with, though not always, doubtful vested interests, links of patronage and possibly, leakages. Large scale construction works offer more possibilities of malpractices than do small works, as is hinted by experienced officials.

Second, and more important, is the magnitude of the tasks. As seen in the field, the canal irrigation systems which, for many years have been the source of relative prosperity, are today the main obstacle to future growth. To reach average yields of 5000 kg./ha. of paddy per crop in the canal systems along the Bay of Bengal, from Orissa to the extreme south, water delivery must be quasi-perfect. Natural and unavoidable constraints due to the monsoon can be reduced by greater efficiency in the use and distribution of water, since losses through seepage can amount to 40-50 per cent of the available water. This implies formidable repair and renovation and the construction of thousands of kilometres of field channels in order to replace the field to field system. Lining of certain canals is equally important.

In the north-west, the ancient networks need considerable work beyond just better maintenance. For instance, in the Upper Ganga System, irrigation intensity amounts to 119 in the head reach and 68 in the tail, which shows how uneven is the distribution of water and/or the amount of losses within the system.[14]

[14] Note of R. Chambers on a seminar held at the Water Resources Development Training Centre at Roorkee, July 1984.

The situation gets worse further east. In Bihar the old Sone canal system (1860) south of the Ganges has fallen into such disrepair that 559,000 irrigated ha. have dwindled to half in the past ten years.[15] The projects begun after Independence (Kosi, Gandak in Bihar) are not operating satisfactorily. Lack of field channels, uneven distribution of water, inadequate repair of canals, delays, late release of water and lack of field levelling, are only a few of the problems.[16]

According to Robert Chambers, the potential created under major and medium canal irrigation is around 30 million ha., only half of which receives 'significant irrigation.' Except in the *warabandi* system of the north-west and in some delta areas, 'the productivity of irrigation water and the equity of its distribution are typically low.'[17]

Another very serious problem has emerged during the past thirty-five years. Until Independence, most canal irrigation systems relied on diversion barrages on rivers. Since then, several large reservoirs with dams have been built : Bhakra in Punjab, Tungabhadra and Nagarjunasagar in the Deccan as well as many other reservoirs. Due to severe erosion, the silting of dams is about to shorten their life span. Only very long term and expansive re-afforestation schemes in the catchment areas will gradually slacken this process.

B.B. Vohra, one of the most experienced civil servants on water management problems, and now Chairman of the Advisory Board on Energy, has been vocal for many years about the need to concentrate on 'maximising the use of existing potential', but he is also concerned with *organisational problems* such as the links between Command Area Development Agencies (CADA) and the irrigation ministry or department in the states. Merged at the centre in some states like UP, they are not always so in other states. The lack of coordination between both agencies is striking in Bihar, while it seems to have been corrected in Andhra. The

[15] *India Today,* 15 April, 1986

[16] Reserve Bank of India, *Report of the Committee on Agricultural Productivity in Eastern India,* vol.I, pp.130-133.

[17] The *warabandi* system means irrigation on a rotation basis with direct access to irrigated plots through field channels. See R. Chambers, To the Hands of the Poor : Water, Trees and Land, paper for the Institute of Economic Growth, Silver Jubilee, Delhi, April 1984.

organisation responsible for land levelling and several other tasks designed to make full use of canal irrigation should work *closely* with irrigation authorities and play an increasing role in order to fill the gap between the irrigation potential created and the actual utilisation. In 1985, the gap amounted to 7 million ha. Two years later the situation was still worse. There is an 'increasing trend of unutilised irrigation potential.'[18]

Taking into consideration these various problems, B.B. Vohra deplores that the Seventh Plan allocates Rs. 111.56 billion for the creation of additional potential and only 16.7 billion for command area development programmes.[19]

Other measures are needed to improve canal irrigation. Water charges are usually so low that many systems operate at a loss. Besides, cheap canal water does not induce farmers to seek complementary irrigation with private tubewells, as observed in parts of south India.

The overall administration of irrigation departments would need drastic overhauling in order to remove deep and long standing maladies. It is sad to see how these departments, prestigious during British days have deteriorated in efficiency and integrity. Many reports refer to these defects, some of which are quoted in R. Chambers' book *Canal Irrigation for People*.[20] Others are plainly stated by B.B. Vohra in his address in Bangalore.

Over and above all these problems are the water disputes between various states. Many projects have been delayed for years and several issues remain unsolved like the dispute between Karnataka and Tamil Nadu on the sharing of waters from the Cauvery basin, the upper reaches situated in Karnataka.

One moves now to another issue, the object of debate for at least two decades : the respective importance of major and medium irrigation as against minor irrigation.

For 1985-86, the potential is anticipated at 69.7 million ha. versus 62.3 million utilised. For 1986-87 the targets are 72.2 and 64.1 million ha. respectively.

[18] *Economic Survey 1986-97*, (New Delhi, 1987)

[19] B.B. Vohra, Towards a National Water Policy, Keynote address delivered at the National Seminar on Command Area Development, Bangalore, December 1985.

[20] Petty corruption was always there, but it was more rare at the higher echelons of the administration.

Gross irrigation potential and utilisation (million ha.)

	1950-51		1984-85	
	Potential created	Utilisation	Potential created	Utilisation
1. Surface water	16.1	16.1	40.2	34.1
a. major,medium	9.7	9.7	30.5	25.3
b. minor	6.4	6.4	9.7	9.0
2. Ground water minor	6.5	6.5	27.6	26.1
	22.6	22.6	67.0	60.4

These data confirm the growing role played by ground water irrigation (tubewells, open wells with pumpset). The pace of progress in major and medium irrigation has been much slower and much more costly. Poor surveys, defects in planning and shortage of funds explain delays in construction. However, to overcome these long standing and well recognised shortcomings seems an onerous task. In spite of its emphasis on curbing these problems, the Sixth Plan has not been too successful. Sixty-five major projects started before April 1976 should have been completed during the Plan period. Only twenty-five have been completed so far (*Seventh Plan*, vol II, p. 73).

Cost escalation is even more worrying. B.B Vohra quotes a study which shows that over the years, 156 major irrigation projects are now expected to cost Rs. 140 billion versus an original estimate of Rs. 21 billion.[21]

While the criticisms against these delays and rising costs are legitimate, I wonder whether there has not been a tendency to grossly underestimate the complexity of the tasks involved, the long duration needed not only to complete the basic irrigation works but also to achieve the tasks covered by CADA. Viewed from that perspective, past mistakes are perhaps more explicable, a matter not confined to India, but found also in China, Pakistan and other Asian countries.

Should one, as is often suggested, reduce investments in major works in favour of ground water irrigation?

During the Seventh Plan, the cost of creating new potential in major-medium irrigation amounts to Rs. 27,000 per ha., to which are added Rs. 13,000 per ha. for the cost of command area

[21] B.B. Vohra, Towards a National Water Policy, Bangalore, December 1985.

development and drainage.[22] On the other hand the investment cost of irrigation by tubewells amounts to Rs. 3,000-5,000 in the plains.

Wherever minor projects of ground water irrigation are possible, one should obviously opt for it, but there are large areas unfit for such massive use of tubewells. Taking examples like the Rajasthan Canal or Nagarjunasagar, there was no other solution than works of that magnitude.

Coming to tubewells and pumpsets, electricity has been a major incentive for farmers, but today it creates serious difficulties. As seen in western UP, tubewells cannot operate for sufficiently long hours to replace low yield coarse grains in the monsoon by paddy which requires more irrigation, as is done in Punjab. Further east it is the same story. In eastern UP wide tracts of paddy cannot be irrigated in the monsoon, so that crops remain poor, which is only partly compensated by better wheat crops under irrigation.[23] The situation gets worse in Bihar. These troubles are not lacking in the south either despite the fact that state electricity boards are more efficient. As seen in Thanjavur, the supply of power is not very reliable.

Although the Seventh Plan, like the previous one, gives emphasis to electricity, the gap between supply and demand may, at best, be only reduced. Why not then expand oil engines for pumps since they are already in use in certain areas? First, only relatively rich farmers—as seen in Mirpur—can afford to have an extra oil engine to use when power fails. Second, all the small farmers who have installed an electric tubewell or pumpset on a well cannot afford to suddenly replace their electric equipment with an oil engine. Third, the cost of oil tubewells is higher than electric ones and more cumbersome. It is no mean task to bring oil to so many tubewells.

One does not of course, deny that there is room for manoeuver in encouraging the use of oil—which is already practised on a fairly large scale. But it may be more complex than relying on electricity. In other words, while power has been a big incentive and a major factor of growth in the early stages, it has now become a major obstacle to further progress.

[22] B.B. Vohra, Seminar on Surface and Ground-water Resources, New Delhi, GWB/UNDP, February 1986.

[23] Wheat requires much less water than rice, so that irrigation is less inadequate.

Another long standing defect of irrigation refers to the decay of tank irrigation in the Deccan, a matter brought to my notice whenever I visited Andhra and Tamil Nadu beginning with the 1960s. As a result of the neglect of repair and maintenance over the years, desilting operations have become so expensive that, for shortage of funds, the limited attempts made to improve their capacity lay more emphasis on strengthening the dykes and improving sluices. Now in Tamil Nadu alone, tank irrigation covers 30 per cent of the total irrigated area.[24]

Even though improving the existing irrigation system presents a formidable task, one must consider the scope for further irrigation.

	Ultimate potential (million ha.)
1. Surface water	73.5
a. major, medium	58.5
b. minor	15.0
2. Groundwater	40.0
	113.5

Source: Government of India, *Seventh Five Year Plan* (New Delhi, Planning Commission).

The potential created (1951-1985) by major and medium irrigation involved Rs. 152 billion. The additional potential contemplated in the Seventh Plan amounts to 4.3 million ha. requiring Rs. 115 billion. When one adds the losses in maintenance and operations due to inefficiency and low water charges, one wonders as does B.B Vohra, whether India will be able to sustain such rising costs for very long.

All these complex irrigation problems should not make us forget other water management problems. While waterlogging and salinity are not as widespread as in Pakistan, they do necessitate more investments than now. The total area affected amounts to 6 million ha., which needs drainage and flushing of salt.

Flood control and drainage represent more serious issues. The former may affect up to 40 million ha. out of which 32 million ha. could be protected. By March 1985, 13 million ha. had been protected through 12,531 km. of embankments, 26,000 km. of drainage

[24] See the very lucid study by S. Guhan, *Irrigation in Tamil Nadu*, (Madras Institute of Development Studies, 1984), p. 72.

channels, protection works in cities and villages raised to a higher level to be less exposed to floods.[25] In a very interesting article, the French geographer Jean Racine points out that between 1953 and 1980, the average cultivated area affected by floods has been 3.6 million ha. In the same article, the author quotes several Indian sources criticising political interference leading to malpractices in flood control works.[26]

Flood control works remain important in spite of various shortcomings but they proceed more slowly in areas affected even in normal years by excess water which does not prevent crops to be raised, but considerably reduces their yields. 'Drainage has not received adequate attention due to the obsession with the creation of new irrigation potential,' write the authors of the Reserve Bank Report on Eastern India.[27] In the meetings I had in state capitals, I got the same impression. Drainage does not catch the imagination of politicians like new irrigation projects and the limited funds allotted for that purpose are often cut under budget constraints.

If we try to sum up all these problems—improvement of existing irrigation systems, a key issue to boost yields and multi-cropping further; the future expansion of irrigation with unbearable costs in major irrigation; severe constraints of electricity for tubewells, flood control and drainage—the magnitude of the tasks and their complexity looks really formidable.

Is there enough awareness among the political leadership in the states at the levels of policy and planning of the crucial need for more efficiency, more imagination and a strong determination 'to have things done' and more or less in time? Is there also enough determination, ask several Indians, to fight complacency, leakages or plain corruption which also affect the implementation of projects?

All these problems are handicapped by seeming to be too technical for the layman, but to the farmer they are absolutely essential. A number of senior officers in government and some enlightened politicians have analysed these issues in depth, but one would

[25] Government of India, *Annual Report 1984-85* (New Delhi, Ministry of Irrigation).

[26] J. Racine, Climat et politique en Union Indienne, *Herodote*, no. 39.

[27] Reserve Bank of India, *op. cit.*, vol. 1, p. 72.

need to conduct the water management policy on a war footing if one wishes to quickly eradicate long standing defects and improve future expansion.

The importance of rainfed agriculture

At present rainfed areas supply 40 per cent of foodgrain and the largest proportion of pulses, oilseeds and cotton. Rainfed agriculture will continue to be important regardless of the expansion of irrigation in future and assuming that the full potential of irrigation will be tapped during the next twenty-five years.

	1985	*2010*
Gross cropped area	175	210 million ha.
Gross irrigated area	60	113

Source: *Seventh Five Year Plan, op.cit.*

From a national perspective the progress of irrigation could supply enough additional foodgrain to cover the needs of the population taken as a whole without resorting to imports, except perhaps, in case of extremely severe natural calamities. But as noted in the field, this is only part of the problem. Agriculture must also grow in order to create more local income and employment for all the people who will not be absorbed in large cities, even in industrially advanced states like Maharashtra and Gujarat, and more so in isolated regions with limited outlets like the plateaus of Bihar and wide tracts further south in the Deccan.

Can one say that rainfed agriculture has been neglected until recently? I do not subscribe to this frequently heard thesis because, in the 1960s it would have been suicidal to lay the major emphasis on unavoidably slow growth regions while quicker results could be obtained in irrigated plains areas. Today, when the food balance sheet at the national level looks more promising it is perfectly legitimate, while still pursuing efforts in irrigation, to devote more human and financial resources to rainfed agriculture as the Seventh Plan is doing. In the states, one is more concerned and research is more actively pursued.[28] In New Delhi, more work is

[28] In addition to *ICRISAT,* I would like to mention the work of the Indian Council for Agricultural Research (ICAR) and the All India Crop Research Project on dry land also in Hyderabad, not to mention a number of small research centres as the one visited in Jodhpur and also, CAZRI.

being done under the very able and dynamic guidance of the agricultural commissioner for dry land, who started his career with successful field work in Karnataka.

As Dr. N.S. Jodha, an eminent specialist of dry land with ICRISAT, stated in one of his papers : 'most of the technological options for dry land agriculture are yet in the pipe-line stage.' I had a similar impression after visits to several research centres and discussions with experts. The dominant impression is one of a great complexity of solutions, compared to irrigated crops, which enjoy 'homogenising factor.' In rainfed agriculture, climate and soil differences lead to a lot of intra-regional heterogeneity. 'Therefore,' adds N.S. Jodha, 'available dry farming technologies continue to be inapplicable for several areas', hence the need for considerable more research.[29] For instance, in the penisular plateau of the north-east, low-lying lands may receive too much water, while uplands are covered with light soils where the run-off of water and erosion are high. In both cases higher yields are thus difficult to achieve. Again in low-lying lands broadcast paddy is sown due to excess water. Only local varieties can grow because new seeds would grow far too slowly and may be overtaken by weeds more easily than local varieties.

Problems are even more tricky when tackling black soils, so heavy and sticky in the monsoon that it is better to grow a *rabi* crop from August to December (or a bit later), and light red soils only fit for monsoon crops.

The Bombay Dry Farming system relying on soil moisture conservation through contour bunding was not successful. Water tended to accumulate along the bunds and 'due to the absence of alignment between land contours and property boundaries on land, farmers resisted the contour bunding'.[30]

Today the emphasis is on light ridges combined with small ditches, broad bed and furrow system to increase soil moisture by checking the run-off of water and reduce erosion. Inter planting, an old practice, is encouraged through a number of combinations: *jowar* or soyabean and some pulses like pigeon pea.

[29] N.S. Jodha, Development Strategy for Rainfed Agricultural Possibilities and Constraints, paper presented at the Institute of Economic Growth, Delhi, 27 to 30 April, 1984.

[30] *Ibid.*

Through the watershed approach, water is collected in a tank which improves ground water supply or can be used directly for light irrigation. One can harvest around at least 1,000 kg./ha. of *jowar* and 500 kg. of pulses, the latter enriching the soil by fixing nitrogen from the air..

The implementation of the watershed technology on a fairly large scale may not always prove successful. Experiments in Andhra and Maharashtra brought the farmers a net profit not much higher than that from their traditional plots, because of 'intensive use of inputs, especially fertilisers.' Elsewhere, the broad bed and furrow technique did not prove much superior to flat cultivation.[31]

Studying these problems in Gulbarga district (Andhra), Ruedi Baumgartner also observed several snags in the way of the new dry land techniques.[32] The watershed approach requires group action because even on a small watershed there are several plots belonging to different owners which can range between 7 ha. to 93. ha. according to some surveys.[33]

Other experts are of the opinion that the watershed approach and even small ditches are not necessary. Besides, in cases of severe storms, the latter would not control water and, as a whole, simpler solutions could be selected. R.G. Grimshaw, an experienced officer of the World Bank in New Delhi, advocates (instead of soil conservation measures), 'vegetative soil conservation measures' based on contour furrow planting and interplanting of a grass species *Vetiveria zizanioides* or *khus* that grows wild in India. With its deep roots, it protects the soil. It is not eaten by cows and can be used for making paper. Already in operation in Nigeria, this method can lead to grain yields of 2,000 kg./ha. with a rainfall of 500-600 mm.[34]

After soil conservation comes agronomic research, more strongly encouraged since the 1970s. New short duration varieties of pulses are experimented with, often with inter-cropping.

[31] See R. Sarin and J.G. Ryan, *Economic Assessment of Improved Watershed Based Technology Options in On-Farm Experiments*, (Hyderabad, ICRISAT, 1983).

[32] Personnal communication.

[33] N.S. Jodha, 'Some Dimensions of Traditional Farming System in Semi-Arid Tropical India'.

[34] Information supplied by R.G. Grimshaw, World Bank Office, New Delhi.

Scientists are working on new varieties of oilseeds, but so far one cannot talk of a genetic breakthrough. One must remember that the breakthrough in cereals was like the tip of an iceberg. It relied on over a century of research in wheat and maize conducted in temperate countries and then adapted to semi-tropical conditions of Mexico. For rice, research had begun vigorously in Japan by the 1920s. Now, when starting from zero or nearly so, as in pulses and oilseeds, progress is bound to take time.

Hybrid *jowar* and *bajra* varieties have been released since the mid–1960s and have shown some success. However, according to N.S. Jodha, 'the hybrids of these crops could spread only in limited areas'.[35] Hybrid *jowar* is not suitable for the *rabi*. Besides, in many areas, risks are too great to enable farmers to use chemical fertilisers.[36]

As far as oilseeds are concerned, another way out lies with the introduction of new plants. The success of soyabean in Madhya Pradesh is encouraging. Starting with 32,000 ha. in 1971, the area reached one million ha. in 1984 and a production of 730,000 t. for the whole country. Sunflower fields can be seen when driving through Karnataka and Maharashtra. One must also realise that traditional oilseeds often grow on poor, if not marginal land and are severely exposed to 'major pest and disease attack.'[37]

For a number of years, the bid to replace rainfed paddy with maize has been advocated, which could bring better yields, but the alternative is not without risks. Maize can suffer from severe mildew attacks. Now, more resistant new varieties have appeared but they must be renewed every four years.

When tabulating all these factors, it is difficult to envisage a rapid acceleration in the growth of production. We have also seen that the Seventh Plan is conservative in its targets for coarse grains, whereas it seems rather too optimistic with regard to oil seeds, expecting as it does an increase of nearly 50 per cent.

The biggest hurdle is, that due to the lack of a good and assured

[35] N.S. Jodha, Some Dimensions of Traditional Farming in Semi-Arid Tropical India, *op. cit.*

[36] Unlike high yield varieties of wheat and rice, hybrid seeds must be renewed every year, which increases production costs.

[37] *Evaluation Report on Oilseeds Development Programmes 1976-80*, New Delhi, Programme Evaluation Organisation, Sept. 1981.

water supply, the return of chemical fertilisers is low or even negative, which prevents a real breakthrough in yields.

On the other hand, it is true that even a moderate increase in yields can boost production in view of the wide areas considered. At present 37.5 million ha. receive less than 750 mm. of water per year with a yield of 400 kg./ha. of grain. Only one extra quintal would mean nearly 4 million t. of grain. It is also true that even coarse grain yields are no longer static and have risen slightly in spite of the irregularity of rains. However, one is a bit sceptical when reading that doubling dry land production is 'a distinct possibility and a realistic goal'.[38] May be, but in the distant future which could be much beyond the year 2010. After what has been seen of the magnitude of soil deterioration, to merely improve it on such vast areas will take a very long time.

In rainfed areas, the increase in production is unavoidably slow, while in irrigated areas there were dramatic increases in yields over a brief period of time.

Other related issues are no less problematic, such as afforestation which could help soil conservation tremendously. We have seen that eucalyptus plantations, small and large, are expanding. Now they have become the focus of controversy. According to some scholars the species of eucalyptus grown in India is 'ecologically inappropriate for vulnerable dry land ecosystems.'[39] In other cases as in Karnataka, it has taken away land devoted to fodder, thus increasing the already existing shortage. I cannot pass judgement on these contradictory arguments, but once again, we are faced with the complexity of these problems.

Traditional agriculture of the adivasis

The *adivasis* (Scheduled Tribes) numbered 51.6 million in 1981. Many of them are in the process of being assimilated within the larger Hindu civilisation and, as seen in Deoghar and Keonjhar, follow the same agricultural practices as other farmers. On the other hand, in the vast forest and hill areas around the plains of

[38] All India Coordinated Research Project, *Dryland Agriculture Research in India: Thrust in the Eighties* (Hyderabad 1983), p.2.

[39] See K.S. Dakshina Murthy, Politics of Environment, *Economic and Political Weekly*, 3 May, 1986.

the Brahmaputra, along the Eastern Ghats and parts of central India, isolated tribes have kept their traditions, worship the forces of nature and practice shifting and burning cultivation. Such a technique is by no means 'primitive.' It has its own rationality, but it begins to collapse when, under population pressure, the period of fallow (originally ten to twenty years) begins to shrink and causes soil losses.

Two sets of interconnected problems—technical and socio-political—have gradually appeared. According to the *Seventh Plan* 'A viable solution to the problem of shifting cultivation has yet to be found.'[40] In 1979, I could see projects of mango plantations in Phulbani district (Orissa) which looked promising, but finally did not succeed too well. Technical extension, credit factors and the isolation of several tribes came in the way.

Other difficulties are of a socio-politcal nature. The uprisings of the Nagas and the Mizos in the north-east have been going on for years, although an agreement was reached at last with the Mizos in 1986. Other troubles have started or could occur along the Eastern Ghats. There are already naxalite pockets in Andhra and Vidarbha (Maharashtra). Highly respected personalities like Baba Amte whom I called on in 1984 in Vidarbha, are extremely concerned about the growing tensions between neighbouring Hindu farmers and *adivasis*, as well as their victimisation and other abuses committed by small officials, including the forest department.

As mentioned in chapters 10 and 11, some tribes are on the verge of extinction. Others confront all sorts of difficulties in finding a new way of life. This is a vast field where more has to be done in a very complex social and techno-economic environment. What is needed are technical persons and officers of a high calibre who would show the proper respect due to *adivasi* customs.

Price policies, storage policies and grain distribution[41]

The growing interdependence of agriculture, industry and infrastructure is accompanied by a growing monetarisation of the

[40] *Seventh Plan, op. cit.*

[41] See the important study by P.S. George, *Some Aspects of Procurement and Distribution of Foodgrain in India* (Washington DC, IFPRI 1985), which was a great help to me while I was writing the following pages.

rural economy. Farmers rely more and more on inputs which they pay for, they also sell more of their production, at least in advanced areas. As a result, prices of inputs and of output play an increasing role in the economy.

While government intervention in prices in India goes back to the Second World War, the question began to be tackled in a more systematic manner with the creation of the Agricultural Prices Commission in the mid-1960s, to advise the government on the price policy of agricultural commodities, mainly cereals, while paying due attention to the consumers and producers. At the same time, public procurement and distribution of food, begun during the Second World War, was considerably developed. It was accompanied by a whole programme of constructing government silos for the storage of grain.

Three basic goals are pursued : to set a price policy which can act as an incentive for farmers to produce more; at the same time through public distribution to have an impact on free market prices in order to avoid very severe price variations and to enjoy buffer stocks in case of bad years.

The Food Corporation of India is handling procurement, imports, storage and distribution through a network across the country. Some state governments have also organised their own agencies.

With the end of imports, government procurement now relies only on grain bought locally. Part of it is kept as reserves and the remaining part is sold through fair price or ration shops—private retailers or cooperative—the number of which has risen from 120,000 to 290,000 for the period 1971-1983.

June 1985	: stocks available	29.2 million t.	
1984-85	: procurements	19.2 million t.	{ 9.9 rice 9.3 wheat
December 1986	: stocks available	23.6 million t.	

The minimum stocks considered necessary amounts to 20 million t. Thus, in spite of a rising number of storage facilities like silos, grain has also to be kept on platforms under plastic sheets.

Compared to the total production of foodgrain (around 150 million t.), the amount procured by the state is small, but as a percentage of marketed grain it is quite high corresponding to 35-40 per cent rice and 60 per cent for wheat.

Percentage share of the main states in total procurement

	wheat		*rice*	
	1980-81	*1983-84*	*1980-81*	*1983-84*
Punjab	66.6	62.3	45.4	44.0
Haryana	15.6	16.9	12.1	8.8
Uttar Pradesh	8.1	17.4	10.6	10.4
Andhra Pradesh	–	–	12.3	18.6
Tamil Nadu	–	–	3.1	8.2

Source: Bulletin of Food Statistics.

Here is further evidence of the highly uneven progress in agriculture and we must remember that Haryana and Punjab produced hardly any rice twenty years ago!

Although farmers grumble — as they do anywhere — about prices, be it the procurement price of grain[42] or the purchasing price of chemical fertilisers, the system has clearly contributed to overall development. The production of foodgrain has reacted on the whole in a positive way and fertiliser consumption has grown. Excess price hikes on the free market have been prevented with the help of public distribution at a fair price which had some obvious social value, especially for town dwellers.

The storage policy has met with success, and even too much success, in view of the recent difficulties in storing grain. This is a considerable achievement compared to the tragic situation in the mid 1960s when India had to face extremely severe droughts with no stocks at hand and needed considerable external, mostly American, food aid.

While showing such results, this policy has created its own problems. The cost of storage is becoming very high and one hears conflicting views on storage losses, a point I was unable to clarify. Officials interviewed said that losses remain within acceptable limits, including storage in the open under plastic sheets. Other sources claimed higher losses.

And this is not all. In most states, electricity is subsidised and water rates for canal irrigation are often too low to cover the operation costs of the canals. Then, in certain states there are special schemes for food distribution : free meals in schools in Tamil Nadu, supply of rice at Rs. 2 per kg. for poor people in Andhra, along with other schemes like housing, etc.

[42] The price is reconsidered every year and announced before the harvest.

Central government subsidies (billion rupees)

	Food	*Fertilisers*	*Total*
1970-71	0.18	–	0.18
1975-76	2.50	–	2.50
1980-81	6.50	4.66	11.16
1984-85	11.00	18.32	29.32
1985-86	16.50	14.50	31.00

Source: Official data quoted from B.M. Bhatia, *Indian Agriculture* (New Delhi, Centre for Policy Research, 1985), p. 118.

Both at the central and state level, many voices, B.M. Bhatia's for example, are expressing their anxiety and demanding a curb on subsidies. While a subsidy on food seems to be difficult to reduce, there is perhaps more scope with regard to chemical fertilisers, where a price increase was introduced in February 1986.[43] Because prices are so weighted with politics, no instant and perfect solution is in sight. India, like most other Asian countries must learn to live with them as Western coutries or Japan are doing!

Industrial support to agriculture

Industry and modern infrastructure are increasingly involved in rural development and agriculture.

	Production 1984-85	*Target* 1989-90	
Distribution of certified seeds	7.04	11.74	million qt.
Chemical fertiliser consumption (nutrients)	8.37	13.50–14.0	million t. nutrient
Production of chemical fertilisers	5.18	8.75[44]	million t. nutrient
Pesticide consumption (technical grade material)	50.00	75.00	'000 t.

Source: *Seventh Plan, op. cit.*, vol 2, p. 13.

[43] According to G.M. Desai, the increase in fertiliser consumption owes more to the profitability of fertilisers coupled with other technical innovations than to subsidy. 'Policy for Growth in Fertiliser Consumption', *Economic and Political Weekly*, 24 May, 1986.

[44] Potash is imported for lack of available resources, as well as part of nitrogen and phosphatic fertilisers.

These figures imply continuous improvements with the breeding of new seeds and their multiplication is no mean task, because though seed renewal is progressing, it needs to expand more systematically. For instance, it is distressing to learn that in Punjab, seed renewal is not adequate which, unless quickly corrected, will reduce productivity.[45] This is due more to declining extension services than to political unrest. The production of chemical fertilisers has not kept pace with plan targets until now for various reasons: power shortage, planning difficulties, technical problems leading to construction delays and operations below capacity. The task is two-fold : to renovate several existing units and to create new ones. Even if the targets of the present plan are achieved, a substantial gap will still have to be filled by imports.

Among the key sectors, electricity features prominently. The shortage does not only hinder agriculture with reference to tubewells and small rural industries. The organised sector of industry suffers the most, where shortage of power is one of the major factors explaining why so many plants operate below their capacity.[46]

These constraints came to be felt about fifteen years ago and they have grown despite considerable increase in production. The capacity of power plants has risen by 49 per cent during the Sixth Plan. The present plan also envisages an expansion.

	Installed capacity (MW)
1984-85	42,490
1989-90	64,736

While the shortage amounted to 12.6 per cent in 1980-81, it fell to 6.1 per cent in 1984-85 and should more or less disappear if the Seventh Plan target is reached. Judging from previous failures in achieving targets, it is far from certain that the situation will be normal by 1990. Not only are new plants needed, but the maintenance and operations of the existing power plants and the network of distribution need considerable improvement. There has

[45] New seeds need continuous renewal every four to five years otherwise they deteriorate.

[46] According to estimates of the Federation of Indian Chambers of Commerce and Industry, a 10 per cent power cut costs the economy Rs. 70 billion in production losses.

been some recent progress in the operation of the state electricity boards but the progress has been uneven with many of them working at a loss. Technical weaknesses accompany political interference, malpractices or leakages.

We come then to the question of resource allocation, a problem rather similar to what has been observed for irrigation: i.e., an old established practice to favour new investments at the cost of maintenance and operations. The Seventh Plan rightly insists on improving and modernising the existing network, but from what I gathered in several states, I wonder whether or not the changes in resource allocation go far enough.[47]

To conclude, although an awareness of the absolutely crucial role of power and its economic and social value is increasing, are the changes contemplated in planning and management drastic enough?

How far can the situation be eased by new or alternative sources of energy? Starting with biogas, I did not notice much progress compared to my 1978-79 survey. However, at the all India level, during 1985-86, biogas plants increased by 200,000 'bringing the total number to 650,000'.[48] Since these plants supply gas mostly for purposes of cooking and lighting they have only a limited effect on the reduction of power shortage. The solar photovoltaic system and solar energy are still too expensive to compete with conventional power. Wind energy too needs considerable technological improvement. This summary of the Seventh Plan confirms the views of other experts according to whom in the near future at least, alternative sources of energy cannot be relied upon, which underlines further the need for greater efficiency in the use of conventional energy.

As regards roads, in spite of the considerable progress achieved so far, many more all weather roads are needed, more transport facilities, both trucks and bullock carts. The impact of the railway should also be mentioned, be it for the transport of grain, chemical fertilisers, steel or cement. Although the situation here is less

[47] It is obvious that in many cases it would be cheaper to improve the existing system than to push too many new projects. This of course does not mean that new investments are not needed. The question is that of the adequate share of investments.

[48] *Economic Survey 1986-87*, p. 27.

problematic than in the power sector, the whole transport system has been under stress for many years. As in other sectors, one must, at the same time, improve or renew existing equipment, road and railway lines, and invest in new equipment and new roads.

And population?

What can one add to the endless debate on population, which began even before Independence? India was the first developing country to start family planning programmes, yet, the fall in the growth rate is slow to come, even if we bear in mind the size of the country, with its 571,000 villages. After each of my visits to India, I find an increased anxiety among policy makers, senior officials and urban elites as a whole. One also feels a growing concern among many people in the villages who see their holdings shrinking.

The Registrar General has presented three hypotheses regarding the population for 2001 : a minimum of 950 million, 990 or 1,050 million. The first one implies a fall in the birth rate from 33 per thousand in 1981 to 21 per thousand and in the death rate from 13 per thousand to 9 per thousand. As against 28 per cent of the eligible couples practising family planning now, 60 per cent would be protected in 2001. The growth rate would be cut by half to fall to one per cent per year.

Can one reach this hypothetical target? The fertility rate is declining in certain states and in many towns. Yet, in other states one does not notice significant progress over the past decade. To use direct pressure is impossible after what happened during the emergency. Could greater incentives—as mentioned in certain states—play a large role? This remains an open question. In any case it will be easier to implement such schemes in cities rather than in villages.

One could intensify the propaganda among women who are, as everywhere else, more open to these issues than are men who do not bear the problems associated with pregnancy and delivery, especially in villages. Could one push more vigorously a larger variety of means instead of concentrating perhaps too much on male and female sterilisation; would it be possible to encourage other means as well? There is no evidence to prove that village women cannot use the pill.[49] It has been quite successful in Indonesian villages and in Bangladesh. In Thailand very aggressive programmes have propagated condoms. Injections like depo-provera could also help.

More efforts should be made to improve the quality of family planning services, especially for sterilisation. Complaints from farmers (some evidence has been given in this book) about poor facilities, rough treatment and lack of attention are frequent.[50]

After concretely seeing the meaning of population pressure in the field, one has to envisage at least 30 per cent increase in population over 1981. Here again, the main concern is not to find enough food for all these people, but to find enough land, to obtain yields in *practically all districts* and to create enough jobs.

To sum up

In already advanced areas where the average yield of wheat is around 2,500-3,000 kg./ha. or more, and clean rice around 3,000-3,500 kg./ha. (4,500-4,700 kg./ha. of paddy), any further increase will be possible provided: water management is carried out smoothly; the application of chemical fertilisers increases and improves in terms of a better mix of nitrogen, phosphate and potash, and, in certain areas new products like zinc; seed renewal occurs regularly every four to five years and especially for monsoon crops; pest control becomes more thorough.

These conditions show that the old 'muddle through system' cannot work any more in such areas, whereas in the eastern plains, with yields as low as they are at present (700-1000 kg./ha. of rice, 1,000-1,500 of paddy) in rainfed fields, considerable progress can occur even with imperfect irrigation, uneven use of new seeds, low doses of chemical fertilisers and insufficient pest control. One could, under such conditions, reach in the first stage, one crop of rice at 2,000 kg./ha. followed by the same amount of wheat, instead of no crop in the *rabi* or a very low yield crop as it is now.

[49] In any case there has been a spectacular increase in the number of people who use the pill: 9.3 million in 1984-85 against 5.5 million in 1985-86, *India Today*, 31 July, 1986.

[50] The article of Joyce Pettigrew, Problems Concerning Tubectomy Operations in Rural Punjab, is painful to read. Paper presented at the Conference of European South Asian Institutes, Sweden, 1983.

In the rainfed tracts of peninsular India, only moderate increases can be expected, in view of the twin demanding tasks of fighting for soil conservation and improvement as well as for increasing yields under complex technical conditions.

Bearing in mind the future population increase, a faster process of rural economic diversification is needed in order to widen the employment market in villages and small rural towns. Several districts are already involved in this process, but this can be expanded in other areas.

Finally, we should consider the question about the direct attack on poverty with special schemes versus the indirect approach to poverty, whereby poverty retreats under the impact of a fast and diversified process of growth. Taking into account political factors, even if the government wanted to give up the target group approach, it could not do so. But is it wise to increase further the funds allotted to such schemes as was the case in the 1986-87 budget and again, in the 1987-88 budget? In view of the leakages and other shortcomings mentioned in this book,[51] and in spite of the attempts to improve the system, it is difficult to give a positive answer to this question

Politics and development strategy

On reaching New Delhi in February 1986 after four months in the country, I found myself in central government offices and the office of the Prime Minister with mixed feelings. What a tremendous task it is to govern such a colossal country, with its incredible combination of people, languages, religions, social and political conflicts and its enormous variety of techno-economic problems. Is it fair, as is so often done in the capital's cocktail parties, to talk about the efficiency, laxity and disorder in the country? True, these weaknesses do exist but, at the same time, one tends to underestimate the degree to which the administration works, one forgets that decisions taken are implemented in countless villages, that currents of change are felt nearly everywhere, and that the country holds together. In spite of corruption, one comes across a number of talented and dedicated people in all spheres: in government

[51] See also the statement of the Minister for Agriculture, *Overseas Hindustan Times*, 30 May 1987.

service, private business, education and research. What a stimulating sight it is to meet great old men like B.K. Nehru, then Governor of Gujarat, L.K. Jha, dealing with economic reforms, my old friend Surendra Pal Singh, M.P. from Bulandshahr, regularly touring his constituency, and how many young district officers working under strains of all kinds. In the state capitals and at the Central Secretariat, one comes across a number of gifted and well trained officers, young men in their thirties, senior ones in their fifties.

Both before and after Independence, the Indian Civil Service and the Indian Administrative Service have been the focus of sharp criticism among Indian scholars and certain, though not all, politicians. The 'colonial' aspect of the system deserves serious qualification. In fact, when observing public administration from Kautilya's days until now, one notices, of course, substantial changes in the spirit and the performance, but there is also a considerable amount of continuity. The Mughals borrowed from the old Hindu kingdoms, the British borrowed from the Mughals, and, also, from the Chinese, in particular the idea of competitive examination. As regards IAS officers, is it really fair to accuse them of a colonial mentality, when a number of them were not even born under the British Raj?

After closely observing many such officers in the districts, in state capitals and in New Delhi over a period of more than three decades, and while not denying serious shortcomings and weaknesses, I feel that the IAS system remains an important asset for India. This becomes even more obvious when one looks at countries which do not enjoy such a system, be it in many parts of Africa or Asia. In a country like Afghanistan for instance, the lack of such a civil service has been, among others, one of the factors which led to the present and tragic situation there.[52]

Coming to the political system, one can say that, compared to many other developing countries, India has been lucky enough to enjoy a relatively strong and stable political leadership from 1947 till today. Jawaharlal Nehru laid the foundation of a new India, introducing a democratic system and a wide expansion of the

[52] I have dealt at greater length with this topic in my paper 'La Fonction publique en Inde, le passé, le présent' in F. Bloch-Lainé and Gilbert Etienne ed., *Servir l'Etat,* special issue of the journal *L'Homme,* Paris, 1987.

economy. To Lal Bahadur Shastri during his brief tenture as Prime Minister goes the credit of the agricultural policy which paved the way for the green revolution. Some people criticise Indira Gandhi for lowering the level of public life, while others praise her for keeping the country together in spite of acute and violent tensions. The truth may lie with both currents of thought, but at least one thing is clear: Indira Gandhi left her son an economy that was in much better shape than the one Lal Bahadur Shastri inherited in 1964 from Nehru.

In spite of this relative stability, at least since the late 1960s, one can sense a deterioration in the political climate with rising corruption, falling standards of politics, the proliferation of neither particularly able, nor honest politicians, rising communal violence excited further by fundamentalist currents under Hindu, Muslim or Sikh colours. These diseases have nothing to do with a particular party. They were present during the Janata government from 1977 to early 1980. They exist in states with or without a Congress government.

A number of respected Indians of all walks of life have for many years expressed their growing anxiety. In 1983, V.K.R.V. Rao, a famous economist and former cabinet minister wanted 'a restoration of the values and norms that prevailed by and large during the days of Gandhi and Nehru ... When the value system is in shambles and there is no attempt to restore ethical and social norms of conduct, there is no hope for the future of the people.' One could quote hundreds of similar statements, by intellectuals, officials and enlightened politicians.[53]

These dangerous currents culminated in the assassination of Indira Gandhi in 1984 followed by the equally shocking slaughter of many Sikhs in Delhi and other cities. Was it the end of new India, foreign observers and some Indians asked themselves?

Within a few days, the country once again demonstrated its resilience, its amazing capacity to absorb even the worst crises and remain on its feet. Once again, the common saying 'we Indians always muddle through,' proved correct

[53] V.K.R.V. Rao, *India's National Income 1950-1980* (New Delhi, Sage, 1983) p.196. See also B.K. Nehru, *Thoughts on our Present Discontents* (New Delhi, 1986) for a detailed analysis of political issues, including corruption. Also see I.G. Patel, On Taking India into the Twenty-first Century, *Modern South Asian Studies*, April 1987.

Rajiv Gandhi arrived to power at a very crucial time. The diseases which affect the country are serious enough but the point of no return whereby one would have given up hope for positive changes has been averted. Second, in spite of the deterioration of the political scene, the economy was in relatively good shape. The overall growth had been satisfactory in the early 1980s, especially when considering the adverse international economic surroundings.

The advent of the new Prime Minister was accompanied by a tremendous wave of enthusiasm all over the country, raising at the same time very high expectations of changes in political life. Rajiv Gandhi started almost immediately to tackle a wide range of issues : specific ones like Punjab and Assam, broader ones like the need to improve his own party. In his inaugural speech at the Congress Centenary Session in Bombay in December 1985, he reviewed the ills afflicting the country : the growth of communalism, black money, 'police who do not uphold the law but shield the guilty,' the 'corrupt ways' and 'the self perpetuating clique' within the Congress Party, the 'lack of public morality,' 'our cumbersome administrative machinery'... and, in the field of education 'teachers who seldom teach and students who seldom learn. As to our anti-poverty programmes (they) have to come out of the grip of bureaucratic sloth and inefficiency.'[54]

Obviously the shock wave which had brought Rajiv Gandhi to power could not last forever. The political climate was bound to get calmer, but what is striking is the gap between the euphoria that prevailed in the early months and the situation two years later. By 1986 one could see signs of dissent within the ranks of the Congress Party and outside. But from the beginning of January 1987, a flood of criticism has appeared in the press on the different occasions: the relations between the Prime Minister and some of his senior government servants; the conflict between the former and the President of the Republic; V.P. Singh's shift from Finance to Defence; the resignation of the latter following sharp criticism by a number of Congress MPs and ministers; and other rumours and accusations referring to large scale corruption in arms deals with foreign countries and allegations about money held abroad by various personalities, some of them allegedly

[54] *Inaugural Speech by Congress President,* Bombay, 28 December 1985.

close to the Prime Minister ... Finally, the Congress party was badly defeated in the state elections in West Bengal. It also lost, though less heavilv, in Kerala and faced a real debacle in Haryana.

It is beyond the scope of this book to deal in detail with these different events. What is important is to assess their impact on the country. Do they imply that Rajiv Gandhi has failed as some Indians now claim? At this stage (June 1987) the crisis is no doubt serious, but it is still possible for Rajiv Gandhi to overcome the present difficulties. We should also bear in mind that the microcosm of Delhi, its atmosphere and the role of some leading English newspapers represent only a small—though important—part of the country.

In a way, one feels more concerned about deeper and more dangerous problems. Communal tensions, growing fundamentalist currents in *all* the main religions of India and weak political leadership in many states, are trends that have been in existence for several years and are not abating even now.

Caste and class struggles, or more simply, the fight to share too small a cake, fuelled by very mediocre politicians have been devastating in Bihar and in some parts of UP. Shiv Sena activities in Maharashtra against people from another state and now against Muslims, and dissensions in Gujarat about reservations for backward classes represent other dangers. And let us not forget that quieter states like Orissa contain the risk of potential clashes between the higher castes and others.

On many occasions, the Prime Minister has expressed his concern over corruption and all kinds of malpractices, evidence of which have been given in this book. It ranges from leakages in the implementation of special schemes for the poor or *baksheesh* to get a government job, to much bigger losses affecting the operation of irrigation departments, large scale new hydraulic works, state electricity boards, etc.

With this socio-political background in mind, how does the economic policy look? Its different sectors are becoming increasingly interdependent : a faster growth in agriculture would boost the demand for industrial goods, but the growth of the latter is closely dependent on industrial inputs—more electricity, more roads. At the same time, external trade is equally important. Larger local supplies of oilseeds would reduce imports and thus enable the imports of equipment needed for modernising factories.

All these links show the importance of reforms which will lead to a more efficient economy, the impact of which will be felt down to the village level.

By the mid-1970s itself thinking and rethinking on planning had begun to take place. Some ministers, several senior government servants and economists were beginning to realise that the country could have done better with a more flexible system with less controls on the private sector, which proved a source of waste and corruption. A greater emphasis on foreign trade was also needed. No less crucial was the need to make the public sector more efficient.[55]

It was disturbing to note, as did the well known economist Jagdish Bhagwati, that, in spite of the rate of savings and investments having doubled since the early 1950s, 'the growth rate remained virtually stagnant' until the early 1980s.[56] Isher Judge Ahluwalia adds: 'There is evidence of increasing inefficiency in the use of *factors* in a number of sectors of the economy.'[57]

Initiated in the last years of Indira Gandhi's term of office the reforms gained momentum under her son. There was a retreat of dogmas and rhetoric and a number of important changes were introduced : fiscal reforms, several de-controls in favour of the private sector, a more outward looking policy vis a vis foreign investments and Indian exports and greater efforts to modernise industrial equipments.

Although encouraging results are emerging, much remains to be done. Moreover, one must account for various obstacles : delays, officials dragging their feet, private businessmen unhappy at the loss of comfortable customs barriers or reluctant to take advantage of the liberalisation of the economy, die-hard politicians regretting socialist slogans.

In sum, India has reached a rather critical stage : on the whole the economy is expanding and the growth rate is now rising faster in spite of political difficulties. Yet, the process of development is

[55] See the important books of L.K. Jha, *Economic Strategy for the 80s*, (New Delhi, Allied Publishers, 1980), and, *Growth, Inflation and Other Issues* (New Delhi, Allied Publishers, 1985).

[56] *India Today*, 15 December, 1986.

[57] I.J. Ahluwalia, *Industrial Growth in India* (Delhi, Oxford University Press, 1985), p. 146.

increasingly demanding in terms of efficiency, especially if one wishes to speed up growth further in order to have greater impact on the poor, particularly through the creation of more employment accompanied by higher wages.

Taking the two sets of problems together — the political and economic — what can one say about the future? India has proved that it can sustain a certain amount of disturbance, weaknesses and other shortcomings, but there are also limits to its resilience, especially when considering how complex the whole process of development has become. If the political scene were to deteriorate further, it could slacken the pace of development and the struggle against poverty.

Conclusion

The reader might ask why I talk of a half won battle instead of a half lost battle, considering that so much poverty remains in India and considering also the chasm between the early expectations and the achievements?

While there have been mistakes and shortcomings in planning, one of the major errors was to expect that the eradication of such acute poverty would take only a few decades. Here again one is confronted with the historical perspective. These expectations were not very realistic, judging from past experience, starting with that of the western countries. They took many decades to wipe out extreme poverty and this under much easier conditions of development than were present in Asian countries. Although it seems like a platitude to emphasise it, the time factor has been grossly underestimated, as well as the magnitude of the tasks involved in promoting development.

While overall economic growth has been substantial, one must ask who benefits from it? Although there was evidence even in the 1970s that the poor were benefiting from the green revolution in terms of wages and employment, it is only now that such facts are recognised among fairly large audiences in India and abroad. It is a little sad to see the amount of time that has been needed to move away from a priori biases or sheer ignorance, if not both.

As far as the direct and indirect attack on poverty is concerned, the latter approach seems safer from experience. That is why the ever increasing funds devoted to the former approach seems open to question!

As we have seen, the political scene and its future are not easy to assess at the moment in view of conflicting currents.

The reader might have sensed a contradiction in my observations on politics as well as on administration, where both positive and negative forces are at work. The retreat of dogmas and rhetoric, a move towards a more down-to-earth approach to development and greater practical experience are all positive factors. On the other hand, while not returning to the issues raised at the end of the last chapter, I would like to mention here another matter of concern. Over the years, the relations between civil servants and politicians have deteriorated. Yes-men and opportunists can be found in the administration and there may be many politicians who are closed to the realities of life, who are averse to hearing unpleasant things and who prefer meek officers to strong personalities.

Can one recapture the spirit that prevailed in the early years after Independence when 'Nehru and Patel encouraged independence of judgment'. The latter used to say to his staff: 'If you do not give your honest opinion for fear that it will displease your minister, please then you had better go...[1]

Can one also think, as in the early years, of a cabinet composed strictly of personalities chosen for their competence?

These attitudes are so widespread in so many countries rich or poor, and not only between politicians and civil servants alone, that I wonder whether such a debilitating climate can really be changed? And yet, what a challenging task it would be for a statesman or anybody enjoying high responsibilities, to recapture or strengthen these old values.

However, in spite of these shortcomings and in spite of corruption, India still has a big asset in its civil service, especially when we compare the situation with a number of other developing countries. The acceleration in the pace of reforms under Rajiv Gandhi is no less encouraging.

For a long time, a number of Indians and foreign observers held gloomy opinions about India's achievements, while overemphasising the negative aspects of development. From the late 1970s, this climate began to change. The image of India abroad ceased to focus on cows and beggars and heralded the fact that India had become the tenth largest industrial power in the world. Oddly enough, this was true just after Independence, while now India ranks twentieth or so![2]

Despite this the growing self confidence prevailing in the country remains well founded, especially if we compare the state of the economy in 1950 with the overall increase in the growth rate in the 1980s.

Though highly welcome, these new trends do not mean that the major difficulties are over. As seen in the previous chapters

[1] See B.B. Mishra, *Government and Bureaucracy in India, 1947-1976* (Delhi, Oxford University Press, 1986), p. 6 and p. 92.

[2] Surendra J. Patel, Planned Development in India, *Mainstream,* Republic Day no. 1, 1980.

and during my discussions with experienced Indians, government servants, economists, businessmen and farmers, the next steps may be more difficult than those that have been covered so far: at any rate they will not be easier.

On the one hand, the experience gathered enables planners to avoid many mistakes committed in the 1950s and 1960s, when the only way was to learn through trial and error. On the other hand, we have seen how the whole process of development has become more tricky and more complex, both within India and in India's economic relations with the outside world. That is why it is not certain that in future, the old 'muddle through system' in politics and in development will work as it has done till now.

India's growing concern for economic reforms leading to greater efficiency and better social justice is by no means unique in Asia. Similar currents have gradually emerged since the late 1970s in many countries regardless of their history, culture and ideology

China under Deng Xiaoping is the most striking example but there are others like Pakistan, Bangladesh and other Southeast Asian countries.

The common starting point leads to a reappraisal of the role of the state. The basic idea revolves around less, but more efficient planning, greater reliance on private initiative, less red-tapism and state control, more outward looking economies, both in terms of joint ventures with foreign companies within the countries concerned, and more export oriented policies, along with Asian multinationals expanding in Asia and in the Western world.

All these currents have little to do with the 'new right' in the USA or in western Europe. First, they started *before* the advent of President Reagan to power. Second, Asian countries face problems different from those in the West. In Asia, there is at one and the same time 'too much state,' as in our countries and, 'not enough state.' Vast spheres of socio-economic life require more laws and rules, or better implementation of existing laws. Child labour, food, drugs or chemical adulteration, overloading of trucks, and many other abuses where people are not properly protected, are only a few instances.

The key issue then, is not for or against state intervention, but how to set up a system where the state does not act as a delaying factor in development but stimulates it. Japan and later,

South Korea and the island of Taiwan, are not illustrative of the success of a *laissez faire* policy. Rather, in those countries, the government has been playing an important role in development, setting broad orientations, taking its share of direct or indirect action, while cooperating smoothly with the private sector. In other words, these countries succeeded in harmonising the role of the state and of the private sector in a flexible manner following the changes occurring in the economy.

These broad principles cannot be implemented in exactly the same manner everywhere in Asia. Historical, cultural and sociopolitical factors vary from one country to another. The calibre or the strength of the political leadership varies no less. Yet, what is important is to see how far reforms are carried out in all these countries.

Another significant feature common to most Asian countries is that, like India, they too are, for the first time in history, free from the danger of devastating famines. Silos are full not only in India, but also in China and Indonesia. Pakistan's agriculture is forging ahead. The Philippines is more or less self-sufficient in rice. There are a few cases like Bangladesh where grain imports remain relatively high. Most countries however, have succeeded in raising foodgrain production so that imports have fallen or even disappeared. They have achieved this along lines similar to those in India—through the progress of irrigation, of new seeds and chemical fertilisers. They have also followed a sensible policy of grain procurement and government storage, so that even Bangladesh enjoys reserves in the event of the monsoon failing.

Such progress is particularly significant when compared to Africa. In the early 1960s, experts claimed that there was no food problem in Africa south of the Sahara, while Asian countries were facing a most gloomy future. Today, several countries south of the Sahara have experienced two devastating famines within the last fifteen years and even in normal years, the foodgrain deficit has been rising.

More practical economic policies, growing self-sufficiency in food and, more outward looking development policies are all encouraging factors which favour growth on the one hand and reduction in poverty on the other, but as in the case of India, much still remains to be done. During my visits to China, Bangladesh, Pakistan and India (between 1982 and 1986), I put the same question to a number of experienced government servants about the future of their agriculture. They all agreed that it was certainly not going to be any easier now than what it had been so far. The same is being said in Indonesia. That is why nothing would be more dangerous than to relax on the food front, complacent in having achieved self-sufficiency in cereal production.

For decades, the question of modernisation or development on the one hand and westernisation on the other, has been the focus of debate. It is becoming increasingly clear that countries like India and others are not just aping the West. It is true that consumerism is growing : the craze for televisions, refrigerators, washing machines and other semi-durable goods is spreading all over Asia, including China and it has begun to be felt down in the villages as well. It is no less true that traditional values are being eroded[3] by the process of development, but does this mean a *thorough* westernisation of minds, of social customs and values?

It does not seem like it. In a recent book,[4] the French sinologist Leon Vandermeersch has shown how the area affected by the Chinese civilisation, which includes Japan and Korea, is adapting many western innovations to its own civilisation, a process which can also be observed in South Asia. Lucian Pye has tackled more or less the same issues in a recent book which deals with the far East, Southeast and South Asia.[5] Pye has given evidence of such adaptations or combinations in showing how farmers adopt some new inputs, while retaining several of their traditional techniques and customs.

Similar combinations appear in other spheres of life, such as industry and business in general. The old idea of the close association between entrepreneurship and western civilisation — if not the Protestant ethic to follow Max Weber — has been broken to pieces by the Japanese, the Chinese, the Indian, the Korean, and the Pakistani businessmen.

What will emerge from all these currents : new and external

[3] The process of erosion in fact started several centuries ago, even before the advent of colonial rule.

[4] L. Vandermeersch, *Le nouveau monde sinisé* (Paris, Presses Universitaires de France, 1986).

[5] L.W. Pye, *Asian Power and Politics: The Cultural Dimensions of Authority* (Cambridge, Mass., Belknap Press, 1985).

trends, borrowed and adapted, ancient but partly transformed, be it in social or technical fields — it is difficult to say. Negative results associated with the undue power of money and the retreat of traditional values are matters of concern. But other changes are leading to a decline in poverty, to a more decent life for a large number of people as can be observed in the new industrial countries of the Far East or as can now be perceived in parts of India. The question, at least for me, is not to pass a value judgement, but to observe such changes.

In conclusion, I wonder whether countries like India may not have entered another phase of development. There is definite hope for more rapid growth and faster reduction of poverty. Despite falling political standards and corruption, I was struck once again during the past few years by the calibre, the talent and dedication of a number of people in influential positions in India and in other countries. I was no less impressed by the ability of so many farmers, small or large, who manage to improve their standard of living.

A number of issues, of growth potentialities, of possibilities of greater social justice, etc., are very clear today. It was not so in the 1950s. They can be solved but they involve such challenges that any complacency, any extreme political turbulence or even the return of dogmas, could bring disaster, or at least slacken and hinder the whole process of development.

Bibliography

Only the most recent publications and a few basic books have been listed below. For older publications, the reader could refer to : Gilbert Etienne, *Studies in Indian Agriculture* (1968) and *India's Changing Rural Scene* (1982).

Ahluwalia, I.J., *Industrial Growth in India* (Delhi, Oxford University Press, 1985).

Alexander, K.C., *Agrarian Tension in Thanjavur* (Hyderabad, National Institute of Community Development, 1975).

Ballabh, V., Adoption of High Yield Varieties of Paddy and Wheat in Flood-Prone and Flood-Free Districts of UP, *Indian Journal of Agricultural Economics*, no.1, 1987.

Bardhan, P. and **A. Rudra,** Terms and Conditions of Sharecropping Contracts, *Journal of Development Studies*, April 1980.

____________, *The Political Economy of Development in India* (Delhi, Oxford University Press, 1984).

____________, *Land, Labor and Rural Poverty: Essays in Development Economics* (Delhi, Oxford University Press, 1984).

Bayly, C.A., *Rulers, Townsmen and Bazars, North Indian Society 1770-1870* (Cambridge, Cambridge University Press, 1983).

Benor, D., J.Q. Harrison and **M. Baxter,** *Agricultural Extension* (Washington DC, World Bank, 1984).

Bernard, J.A., *L'Inde, le pouvior et la Puissance* (Paris, Fayard, 1985).

Beteille, A., *Studies in Agrarian Social Structure* (Delhi, Oxford University Press, 1974).

Bhagwati, J.N., in G.Grossman ed., *Wealth and Poverty* (Oxford, Basil Blackwell, 1985).

____________, and **T.N.Srinivasan,** *India* (New York, Columbia University Press, 1975).

Bhatia, B.M., *Indian Agriculture, A Fifteen Year (1985-2000) Policy Perspective* (New Delhi, Centre for Policy Research, 1985).

Bloch-laine, F. and **G. Etienne** eds., *Servir l'Etat* (Paris, 1987).

Blynn, G., *Agricultural Trends in India 1891-1947* (Philadelphia, University of Pennsylvania Press, 1966).

Bhole, H.G., Small Trade as a Lower Circuit in South India's Rural Economy, paper presented at Heidelberg, 9th European Conference on Modern South Asian Studies, July 1986.

Boserup, E., *The Conditions of Agricultural Growth* (London, Allen and Unwin, 1965).

CAZRI (Central Arid Zone Research Institute), *Population, Land, Crops and Livestock Statistics of Arid Zone of Rajasthan* (Jodhpur, 1983).

____________, *Progress in Arid Zone Research* (Jodhpur, 1983).

Chadha, G.K., *The State and Rural Economic Transformation - The Case of Punjab 1950-85* (New Delhi, Sage, 1986).

Chambard, J.L., *Atlas d'un village indien,* (Paris, Mouton, 1980).

Chambers, R., *Managing Canal Irrigation: Practical Analysis from South Asia* (Delhi, Oxford University Press, 1987).

Commander, S., Colonial Rule and Economic Subordinations, the North Indian Economy in the 19th Century, *Etudes Rurales,* Paris, January - September 1983.

Darling, M., *The Punjab Peasant* (London, Oxford University Press, 1947).

Desai, M., S. Hoeber-Rudolph, A. Rudra eds., *Agrarian Power and Agricultural Productivity in South Asia* (Delhi, Oxford University Press, 1984).

District Statistical Handbook, for Puri 1985, Keonjhar 1982, Guntur 1984-85, Kurnool 1985, and Jodhpur 1982. Brought out by the District Statistical Offices.

Etienne, G., *Studies in Indian Agriculture* (Berkeley, University of California Press; Bombay, Oxford University Press, 1968).

____________, *India's Changing Rural Scene* (Delhi, Oxford University Press, 1982).

____________, *Rural Development in Asia* (New Delhi, Sage, 1985).

Frankel, F.R., *India's Green Revolution* (Princeton, Princeton University Press, 1971).

Frykenberg, R.E., *Guntur District 1788-1848* (Oxford, Clarendon Press, 1965).

Galanter, M., *Competing Equalities - Law and Backward Classes in India* (Berkeley, University of California Press, 1984).

George, P.S., *Some Aspects of Procurement and Distribution of Food Grain in India* (Washington DC, IFPRI, 1985).

Gourou, P.S., *Terres de Bonne Esperance, Le Monde Tropical* (Paris, Plon, 1982).

____________, *Riz et Civilisation* (Paris, Fayard, 1984).

____________, and **G.Etienne** eds., *Des Labours de Cluny a la Revolution Verte* (Paris, Presses Universitaires de France, 1985).

Government of India, *Sixth Five Year Plan 1980-85* (New Delhi, Planning Commission, 1981).

____________, *Seventh Five Year Plan 1985-90* (New Delhi, Planning Commission 1985).

____________, *Economic Survey,* Annual Publication (New Delhi, Ministry of Finance).

____________, *Agricultural Census 1970-71* (New Delhi, Ministry of Agriculture).

____________, *Agricultural Census 1980-81* (New Delhi, Ministry of Agriculture).

____________, *Area and Production in Principal Crops in India,* Annual Report (New Delhi, Ministry of Agriculture).

____________, *Annual Report* (Department of Agriculture and Cooperation).

____________, *Annual Report* (Department of Food).

____________, *Annual Report* (Ministry of Water Resources).

____________, *Abridged Report* (New Delhi, National Commission for Agriculture, 1977).

____________, *New Vistas in Land Reforms,* State Revenue Ministers Conference (New Delhi, National Commission of Agriculture, 1985).

____________, *For the Rural Poor* (D. Bandhyopadhyay) (New Delhi, Ministry of Agriculture, 1985).

Guha, S., *The Agrarian Economy of the Bombay Deccan 1818-1941* (Delhi, Oxford University Press, 1985).

Guhan, S., *Pala Kurichi, a Resurvey,* Working Paper no.42 (Madras Institute of Development Studies, 1983).

________________, *Irrigation in Tamil Nadu* (Madras Institute of Development Studies, 1984).

Hanumantha Rao, C.H., *Technological Change and Distribution of Gains in Indian Agriculture* (New Delhi, Macmillan, 1975).

Harriss, J., *Capitalist and Peasant Farming: Agrarian Structure and Ideology in Northern Tamil Nadu* (Bombay, Oxford University Press, 1982).

Herring, R.J., *Land to the Tiller: The Political Economy of Agrarian Reform in South Asia* (New Haven, Yale University Press, 1983).

ICAR (Indian Council of Agricultural Research), *Dryland Agricultural Research in India: Thrust in the Eighties* (Hyderabad, All India Coordinated Research Project for Dryland Agriculture, 1983).

ICRISAT (International Crops Research Institute for the Semi-Arid Tropics), *Research Highlight 1984* (Patancheru, 1985).

Jha, L.K., *Economic Strategy for the 80s* (New Delhi, Allied Publishers, 1980).

________________, *Growth, Inflation and Other Issues* (New Delhi, Allied Publishers, 1985).

Jodha, N.S., Population Growth and the Decline of Common Property Resources in Rajasthan, *Population and Development Review,* no. 2, June 1985.

________________, Development Strategy for Rainfed Agriculture, Seminar at the Institute of Economic Growth, Delhi University, 27-30 April, 1984.

Kumar, Dharma and **T. Raichaudhuri** eds., *The Cambridge Economic History of India,* 2 vols. (Cambridge, Cambridge University Press, 1982).

Maheshwari, S.R., *Rural Development in India: A Public Policy Approach* (New Delhi, Sage, 1985).

Mellor, J.W. and **G.M. Desai** eds., *Agricultural Change and Rural Poverty* (Delhi, Oxford University Press, 1986).

Mishra, B.B., *Government and Bureaucracy in India, 1947-1976* (Delhi, Oxford University Press, 1986).

________________, *National Seminar on Ground Water Development, A Perspective for Year 2000,* December 1983 (Roorkee, University of Roorkee).

Nehru, B.K., *Thoughts on Our Present Discontents* (New Delhi, Allied Publishers, 1986).

Patel, I.G., On Taking India into the Twenty-first Century, *Modern Asian Studies,* April 1987.

Pye, L.W., *Asian Power and Politics, The Cultural Dimensions of Authority* (Cambridge, Mass., Belknap Press of Harvard University Press, 1985).

Racine, J., Climat et politique en Union Indienne, *Herodote,* no. 39, 1985.

Raj, K.N., N. Bhattacharya, S. Guha, S. Padhi eds., *Essays on the Commercialisation of Indian Agriculture* (Delhi, Oxford University Press, 1985).

Rao, V.K.R.V., *India's National Income 1950-1980* (New Delhi, Sage 1983).

Reddy, V.N., Patterns of Commercial Farming in Guntur District, *Economic and Political Weekly,* 21 December 1986.

Registrar General, *Primary Census Abstract 1981* (New Delhi, 1983).

Reserve Bank of India, *Report of the Committee on Agricultural Productivity in Eastern India,* Bombay 1984, 2 vols. (Chairman S.R. Sen)

Rogers, G., *Poverty and Population: Approaches and Evidence* (Geneva, International Labour Office, 1984).

Roy Chaudhury, P.C., *Darbhanga* (reprints, Bihar District Gazetteers, 1958).
______________, *Muzaffarpur* (reprints, Bihar District Gazetteers, 1964).

Sarin, R. and **J.G. Ryan,** *Economic Assessment of Improved Watershed-based Technology Options* (Patancheru, ICRISAT, 1983).

Schultz, T.W., *Transforming Traditional Agriculture* (New Haven, Yale University Press, 1964).

Sen, B., *The Green Revolution in India: A Perspective* (New York, Wiley, 1974).

Shah, C.H., and **C.N. Vakil,** *Agricultural Development of India* (New Delhi, Orient Longmans, 1979).

Stein, B., *Peasant, State and Society in Medieval South India* (Delhi, Oxford University Press, 1980).

Thorner, D., Coastal Andhra Towards an Affluent Society, *Economic and Political Weekly,* annual no., February 1967.

Vohra, B.B., Towards a National Water Policy, Seminar on Surface and Ground Water, Bangalore, December 1985.

______________, *Land and Water Management Problems in India* (New Delhi, Ministry of Home Affairs, Department of Personnel, 1982 [second ed.]).